5 STEPS TO A

5

500
AP Statistics Questions
to know by test day

Get ready for your AP exam with McGraw-Hill's *5 Steps to a 5: 500 AP Statistics Questions to Know by Test Day*!

Also in the 5 Steps series:

5 Steps to a 5: AP Statistics

Also in the 500 AP Questions to Know by Test Day series:

5 Steps to a 5: 500 AP Biology Questions to Know by Test Day
5 Steps to a 5: 500 AP Calculus Questions to Know by Test Day
5 Steps to a 5: 500 AP Chemistry Questions to Know by Test Day
5 Steps to a 5: 500 AP English Language Questions to Know by Test Day
5 Steps to a 5: 500 AP English Literature Questions to Know by Test Day
5 Steps to a 5: 500 AP Environmental Science Questions to Know by Test Day
5 Steps to a 5: 500 AP European History Questions to Know by Test Day
5 Steps to a 5: 500 AP Human Geography Questions to Know by Test Day
5 Steps to a 5: 500 AP Microeconomics/Macroeconomics Questions to Know by Test Day
5 Steps to a 5: 500 AP Physics Questions to Know by Test Day
5 Steps to a 5: 500 AP Psychology Questions to Know by Test Day
5 Steps to a 5: 500 AP U.S. Government & Politics Questions to Know by Test Day
5 Steps to a 5: 500 AP U.S. History Questions to Know by Test Day
5 Steps to a 5: 500 AP World History Questions to Know by Test Day

5 STEPS TO A 5

500
AP Statistics Questions
to know by test day

Jennifer Phan
Divya Balachadran
Jeremi Ann Walker

New York   Chicago   San Francisco   Lisbon   London   Madrid   Mexico City
Milan   New Delhi   San Juan   Seoul   Singapore   Sydney   Toronto

ISBN 978-07-178070-4
MHID 0-07-178070-X

e-ISBN 978-0-07178071-1
e-MHID 0-07-178071-8

Library of Congress Control Number 2011936619

CONTENTS

ABOUT THE AUTHORS

Jennifer Phan teaches AP Statistics in Los Angeles, CA. She holds a Master's degree in mathematics and a graduate certificate in statistics from the University of Idaho.

Divya Balachadran holds a BS in mathematical statistics from Monash University, Australia. She also holds an MBA in eBusiness from the Monash and teaches at Clark College, WA. She has held varied top management positions in both the public and private sectors.

Jeremi Ann Walker is a mathematics instructor at Moraine Valley Community College. She holds an MS in computational and applied mathematics from the University of Minnesota, Duluth.

INTRODUCTION

Congratulations! You've taken a big step toward AP success by purchasing *5 Steps to a 5: 500 AP Statistics Questions to Know by Test Day*. We are here to help you take the next step and score high on your AP Exam so you can earn college credits and get into the college or university of your choice.

This book gives you 500 AP-style multiple-choice questions that cover all the most essential course material. Each question has a detailed answer explanation. These questions will give you valuable independent practice to supplement your regular textbook and the groundwork you are already doing in your AP classroom. This and the other books in this series were written by expert AP teachers who know your exam inside out and can identify the crucial exam information as well as questions that are most likely to appear on the exam.

You might be the kind of student who takes several AP courses and needs to study extra questions a few weeks before the exam for a final review. Or you might be the kind of student who puts off preparing until the last weeks before the exam. No matter what your preparation style is, you will surely benefit from reviewing these 500 questions, which closely parallel the content, format, and degree of difficulty of the questions on the actual AP exam. These questions and their answer explanations are the ideal last-minute study tool for those final few weeks before the test.

Remember the old saying, "Practice makes perfect." If you practice with all the questions and answers in this book, we are certain you will build the skills and confidence you'll need to do great on the exam. Good luck!

—Editors of McGraw-Hill Education

5 STEPS TO A 5

500

AP Statistics Questions
to know by test day

CHAPTER 1

Overview of Basic Statistics

1. Which of the following is an example of qualitative data?

   (A) Percent of ozone loss
   (B) AKC dog breed
   (C) Class size
   (D) Amount of money spent on advertising during the Super Bowl
   (E) Gross Domestic Product (GDP) percentage

2. Which of the following is an example of quantitative data?

   (A) Radiation levels in millirems of food in Japan
   (B) Fashionable colors by season of the year
   (C) Gender
   (D) High school grade level—freshman, sophomore, junior, or senior
   (E) Favorite sport

3. Which of the following is an example of discrete data?

   (A) Lifetime (in hours) of 35 fluorescent light bulbs
   (B) Weights of dogs (in pounds) at the Seal Beach Animal Care Center
   (C) Temperature (in Fahrenheit) of the Pacific Ocean at Huntington Beach
   (D) Number of hamburgers sold each day at Hamburger Mary's
   (E) Amount of caffeine (in milligrams) for 8 ounces of popular drinks

4. Which of the following is an example of continuous data?

   (A) Number of Netflix DVDs returned each day
   (B) Number of foreclosures in Las Vegas each month
   (C) Value of the New York Stock Exchange (NYSE) composite index each day at closing
   (D) Number of heartbeats per minute
   (E) Number of contacts you have on your cell phone

are used to and
numbers summarize describe
data

5. Choose the statement below that is an example of a descriptive statistic.

    (A) The average broadband speed (mbps) for Internet access in the USA is 4.8.

    (B) "Studies showed glaciers have lost volume on average 'ten to 100 times faster' in the last 30 years" compared to prior time periods.

    (C) A random sample of surnames in Milan, Italy, showed that 32% were Torino. The study concluded that 32% of all people living in Milan have the last name Torino.

    (D) Joe randomly sampled 10 pages of his thesis and counted 6 typographical errors. He concluded that there are 30 typographical errors in his entire 50-page thesis.

    (E) Kaplan's IT department extrapolated from evidence provided by teachers regarding the number of mathematical errors in the curriculum that students' grades should be increased by 4%.

from population to describe and make inferences
random sample data

6. Choose the statement below that describes the use of inferential statistics.

    (A) The average grade on Test 3 in Spring Semester College Algebra was 63%.

    (B) "By assessing the coloration and state of health of 195 free-living urban pigeons, they found that darker pigeons had lower concentrations of a blood parasite called haemosporidian."

    (C) U.S. Department of Education data from 2007–2008 found that KIPP charter schools received $12,731 per student.

    (D) The average age of students in Biology 101 is 23.5 years.

    (E) The standard deviation of the pain scores from the Tylenol treatment group was 1.25.

7. Choose the statement below that describes the use of inferential statistics.

    (A) Tallest student at Saint Anthony's High School, South Huntington, NY

    (B) The proportion of subcompact hybrid cars in the USA reporting over 45 miles per gallon (mpg) in city driving

    (C) The proportion of teachers belonging to a trade union based on a survey done by the National Education Association (NEA)

    (D) Median home price in Long Beach, California, of 50 homes selected randomly from the Multiple Listing Service (MLS)

    (E) Volatility of the New York Stock Exchange

8. Choose the statement below that describes a parameter. *characteristic of population*

    (A) Gallup poll on "life evaluation" shows 53% "thriving" and 44% "struggling."
    (B) Proportion of wrinkled peas in a sample collected in an organic garden
    (C) The standard deviation of the age of students attending community college based on 10 representatives from each campus
    (D) Average salary of all professors at San Francisco City College
    (E) Average concentration of lactic acid in 30 samples of cheddar cheese from Cheddar Gorge, UK

9. Identify the situation where you would conduct an experiment.

    (A) A farmer wants to study the yields of three new varieties of peas developed by Monsanto, Burpee, and Heirloom.
    (B) A poll is conducted to determine the proportion of people cheating on federal income taxes.
    (C) An educator wishes to determine the average IQ of students pursuing a bachelor's degree in mathematics.
    (D) A baseball fan wishes to determine who is the best pitcher in Major League Baseball.
    (E) A doctor wants to maintain a complete record of medical information on each patient.

10. Identify the situation where you would conduct a survey.

    (A) Gallup wishes to determine the proportion of people who see themselves as "thriving."
    (B) Stanford researchers want to determine the effect that improving student vision has on learning.
    (C) A pharmaceutical company wants to advertise that its painkiller is more effective than aspirin, ibuprofen, and Tylenol.
    (D) A gambler wishes to calculate the expected value of buying two lottery tickets.
    (E) The U.S. Army wants to find the average cost of training a cadet.

11. Which of the following describes a random variable?

    (A) The price of a barrel of crude oil on the commodities market on April 1, 1984
    (B) The number rolled on a fair die
    (C) The grade earned on an exam by a student
    (D) The time required by Jaouad Gharib to run a marathon
    (E) The weight of a gallon of water

12. When we collect data for a study, why do we need to take extra care in the data collection process?
    - (A) To get the right average
    - (B) So that we can generalize
    - (C) To avoid bias
    - (D) So we don't make any mistakes
    - (E) To make sure we get the best data

13. Which of the following measures central tendency?
    - (A) Standard deviation
    - (B) Mean and median
    - (C) Interquartile range and range
    - (D) Variance
    - (E) Correlation coefficient

14. Which of the following measures the spread of data?
    - (A) Mode and mean
    - (B) Median and interquartile range
    - (C) First and third quartiles
    - (D) Correlation coefficient
    - (E) Standard deviation and variance

15. Which statement is true?
    - (A) A simple random sample should not be used in an experiment.
    - (B) Inferential statistics are used to estimate population parameters.
    - (C) Parameters are no different from statistics.
    - (D) Descriptive statistics are graphs such as scatter plots or histograms.
    - (E) Regression and correlation are used to describe a single variable.

16. What common sampling technique is the most generally useful for making inferences about a population?
    - (A) Stratified sampling
    - (B) Cluster sampling
    - (C) Random sampling
    - (D) Systematic sampling
    - (E) Representative sampling

17. To study the relationship between two variables, what type of graph would you use?

    (A) Dot plot
    (B) Histogram
    (C) Scatter plot
    (D) Stem plot
    (E) Frequency polygon

18. To identify the shape of univariate data, what type of graph would be the most useful? *is single-variable data*

    (A) Ogive
    (B) Histogram
    (C) Scatter plot
    (D) Pareto chart
    (E) Pie chart

19. Researchers studied the effects that improving vision with eyeglasses had on educational outcomes. They identified 2,069 students who could improve their vision with eyeglasses. 750 were not offered eyeglasses and 1,319 were. Of the 1,319 offered eyeglasses, 928 accepted the eyeglasses. Students who received the eyeglasses scored significantly higher in both math and science. What was the treatment in this study?

    (A) Being offered eyeglasses
    (B) Not receiving eyeglasses
    (C) Scoring significantly higher in math
    (D) Receiving eyeglasses *← control group is whether accept or not.*
    (E) Being identified as a student who could improve

20. A linguist studied conversation styles by gender and age to assess differences between genders. The data were collected by studying videotapes made of "best friends" who were asked to have a conversation together. Which choice best describes the type of study that was conducted?

    (A) Observational
    (B) Experimental
    (C) Poll
    (D) Census
    (E) Survey

One-Variable Data Analysis

21. Describe the shape of the histogram below.

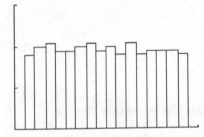

(A) Normal
(B) Bimodal
(C) Skewed right
(D) Skewed left
(E) Uniform

22. Describe the shape of the histogram below.

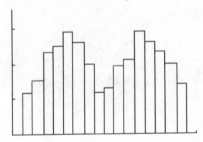

(A) Normal
(B) Bimodal
(C) Skewed right
(D) Skewed left
(E) Uniform

23. Describe the shape of the histogram below.

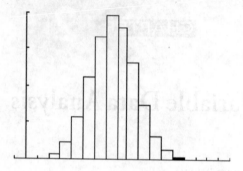

    (A) Bimodal
    (B) Normal
    (C) Skewed right
    (D) Skewed left
    (E) Uniform

24. Find the mode and the median in the dot plot below ($n = 20$).

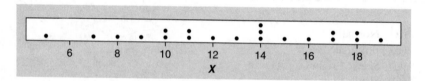

    (A) Mode: 10; median: 13
    (B) Mode: 11; median: 13
    (C) Mode: 14, median: 13.5
    (D) Mode: 14; median: 14
    (E) Mode: none; median: 14

25. Find the modal group and the median in the stem plot below ($n = 20$).
      2   00067889
      3   04499
      4   02399
      5   00

    (A) Modal group: 2; median: 36.5
    (B) Modal group: 3; median: 36.56
    (C) Modal group: 2; median: 32
    (D) Modal group: 2; median: 34
    (E) Modal group: 3; median: 34

26. What are the class boundaries for the bar labeled "10" in the histogram below?

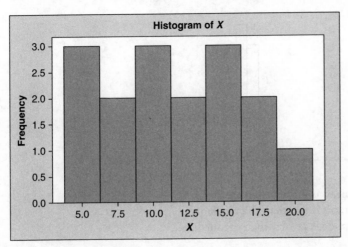

    (A) 7.5–12.5
    (B) 7.25–12.75
    (C) 8.5–11.5
    (D) 8.75–11.75
    (E) 8.75–11.25

27. In the histogram below, which of the following statements is true?

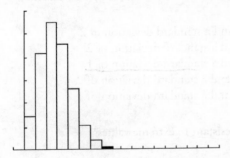

    I. The mean is greater than the median. ✓
    II. The mean is less than the median.
    III. The mode is less than the median.

    (A) I only
    (B) I and III only
    (C) II and III only
    (D) II only
    (E) III only

28. In the histogram below, which of the following statements is true?

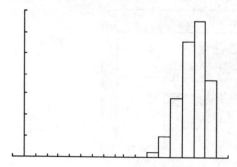

    I. The mean is greater than the median.
   II. The mean is less than the median.
  III. The mode is greater than the median.

(A) I only
(B) I and III only
(C) II and III only
(D) II only
(E) III only

29. In a normally distributed dataset with a mean of 13 and a standard deviation of 2, if the data are standardized by subtracting the mean and dividing by the standard deviation, which of the statements best describes the resulting distribution?

(A) Normal with a mean of 13 and a standard deviation of 2
(B) Normal with a mean of 0 and a standard deviation of 2
(C) Normal with a mean of 0 and a standard deviation of 1
(D) Normal with a mean of 6.5 and a standard deviation of 2
(E) Normal with a mean of 6.5 and a standard deviation of 1

30. What measure of center is most resistant to extreme values?

(A) Mode
(B) Interquartile range
(C) Mean
(D) Median
(E) Range

31. Using this dataset, calculate the mean and the standard deviation.

{3, 8, 10, 3, 12, 7, 10}

(A) Mean is 7.5; standard deviation is 3.5
(B) Mean is 7.6; standard deviation is 3.5
(C) Mean is 7.6; standard deviation is 12.3
(D) Mean is 7.6; standard deviation is 4.0
(E) Mean is 8.0; standard deviation is 4.0

32. Which of the statements is true about the standard deviation?
  I. It depends on the mean. ✓ not only that   measures spread
  II. It is sensitive to the spread. ✓
  III. It is not resistant to extreme values. ✓
(A) I only   ✓ both
(B) I and III only
(C) II and III only
(D) II only
(E) III only

33. Find the five-number summary from the box plot and calculate the interquartile range.

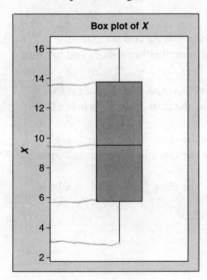

(A) Min: 4; Q1: 5.75; Med: 9.5; Q3: 13.75; Max: 16; IQR: 8
(B) Min: 3; Q1: 5.75; Med: 10.5; Q3: 13.75; Max: 16; IQR: 8
(C) Min: 3; Q1: 5.75; Med: 9.5; Q3: 13.75; Max: 15; IQR: 8
(D) Min: 3; Q1: 5.75; Med: 9.5; Q3: 13.75; Max: 16; IQR: 8
(E) Min: 3; Q1: 5.75; Med: 9.5; Q3: 13.75; Max: 16; IQR: 9

34. Which statement is true of the dataset summarized by this five-number summary?

Minimum	Q1	Median	Q3	Maximum
3.00	4.50	9.00	10.50	20.00

(A) There are no outliers.
(B) 3 is a mild outlier and 20 is an extreme outlier.
(C) 3 and 20 are mild outliers.
(D) 20 is an extreme outlier.
(E) 20 is a <u>mild</u> outlier. There are no extreme outliers.

[handwritten: $1 \times R = 6$, $6 + 1.5 = 9$, 19.5 mild]

35. Which of the statements is true about outliers?
   I. They are natural variation but rare.
   II. They indicate that something may be wrong with the data collection process.
   III. They aren't important and should be identified and then ignored.

(A) I only
(B) I and III only
(C) II and III only
(D) II only
(E) III only

36. Joan's test grade was 84. The class average was 72 and the standard deviation was 4.5. What statement best describes her z-score and her test grade?

(A) $z = -2.67$. Compared to the rest of the class, Joan's grade is low.
(B) $z = -1.67$. Compared to the rest of the class, Joan's grade is a little below average.
(C) $z = 2.67$. Compared to the rest of the class, Joan's grade is high.
(D) $z = 2.67$. Compared to the rest of the class, Joan's grade is a little above average.
(E) $z = 1.67$. Compared to the rest of the class, Joan's grade is a little above average.

[handwritten: $\frac{84-72}{4.5} =$]

37. In 2007 Forest Whitaker won the Best Actor Oscar at age 45 for the movie *The Last King of Scotland*. Helen Mirren won the Best Actress Oscar at age 61 for *The Queen*. The average age for actors is 42.5 with a standard deviation of 7.6. The average age for actresses is 35 with a standard deviation of 9.7. Find the z-scores for each. What statement best describes the results? .3289   2.68

   (A) Whitaker: $z = -0.33$; Mirren: $z = 2.68$. Whitaker's age was about average and Mirren's was well above average.
   (B) Whitaker: $z = 0.33$; Mirren: $z = -2.68$. Whitaker's age was about average and Mirren's was well below average.
   (C) Whitaker: $z = 0.33$; Mirren: $z = 2.68$. Whitaker's age was about average and Mirren's was well above average.
   (D) Whitaker: $z = 2.68$; Mirren: $z = 0.33$. Whitaker's age was well above average and Mirren's was about average.
   (E) Whitaker: $z = 0.33$; Mirren: $z = 2.68$. Whitaker's age was well above average and Mirren's was about average.

38. Use the empirical rule to find the percentage of data that is within one standard deviation from the mean:
   (A) 34%   68-95-99.7
   (B) 68%
   (C) 64%
   (D) 32%
   (E) 65%

39. Use Chebyshev's rule to find the number of data values that are within two standard deviations from the mean if $n = 5,250$:
   (A) 3,900   < 75% 2 std dev
   (B) 3,950
   (C) 3,940   < 89% 3 std dev
   (D) 3,938
   (E) 3,937   3937.5

40. What statement best describes the conditions for using the empirical rule instead of Chebyshev's rule?
   (A) The data are uniformly distributed.
   (B) The data are not normally distributed.
   (C) The data are normally distributed.
   (D) The data are skewed left.
   (E) The data are skewed right.

41. For the probability density curve below, which of the following statements is true?

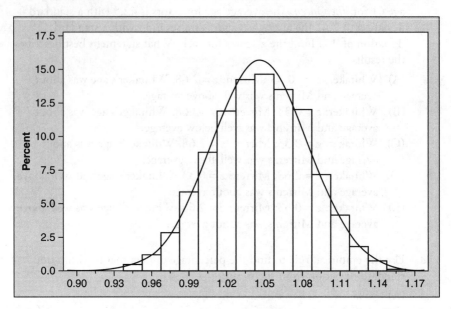

   I. The area is exactly 1 underneath it. *always*
  II. It does not model the distribution of the data. *we don't know*
 III. It is a function that is always positive.

(A) I only
(B) I and III only
(C) II and III only
(D) II only
(E) III only

42. Which of the following is a valid probability density curve?

I.

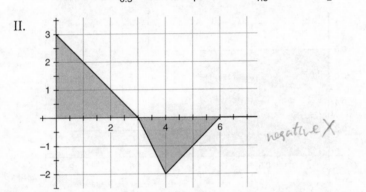

II.

negative X

III.

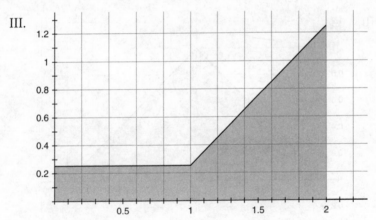

(A) I only
(B) I and III only
(C) II and III only
(D) II only
(E) III only

43. Which of the following is a symmetric probability density curve?

I.

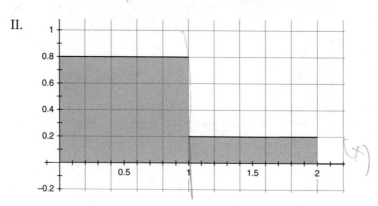

II.

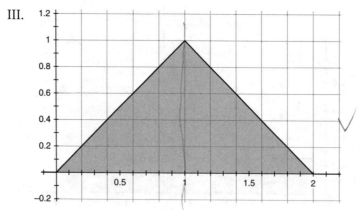

III.

(A) I only
(B) I and III only
(C) II and III only
(D) II only
(E) III only

44. Which of the box plots would match the dot plots shown?

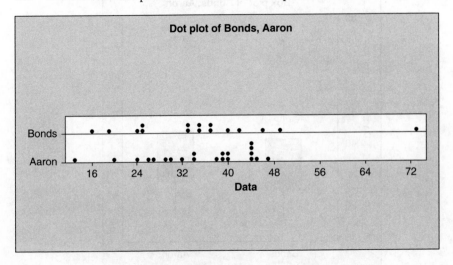

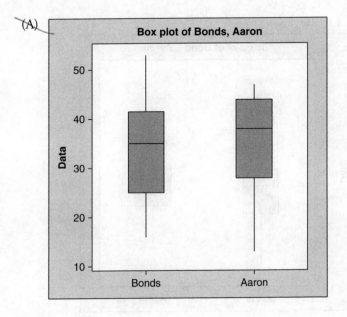

(B)

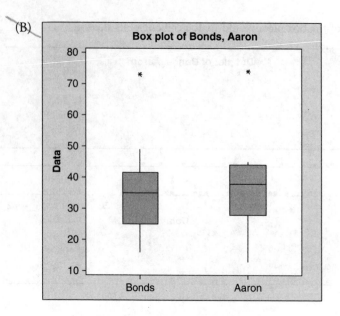

(C)

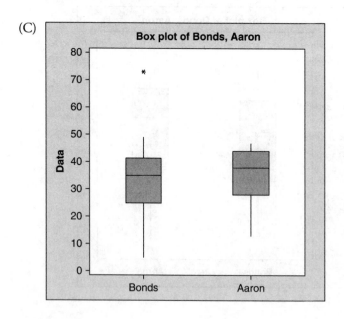

(D)

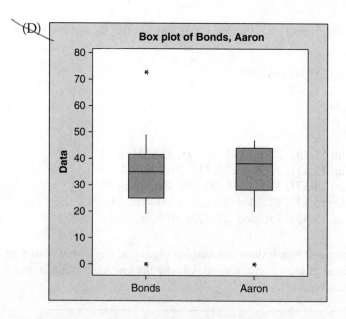

(E)

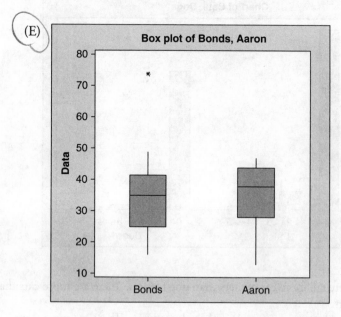

45. Choose the five-number summary that matches this stem-and-leaf plot.

1	3789
2	012333445557889
3	11234578
4	347
5	1

(A) Min: 13; Q1: 22.5; Med: 27; Q3: 33.5; Max: 51
(B) Min: 13; Q1: 23; Med: 26; Q3: 34; Max: 51
(C) Min: 13; Q1: 22.5; Med: 26; Q3: 33.5; Max: 51
(D) Min: 13; Q1: 23; Med: 27; Q3: 33.5; Max: 51
(E) Min: 13; Q1: 23; Med: 27; Q3: 34; Max: 51

46. The histograms below show the number of cat and dog owners based on the number of each animal owned. Choose the **best** description from the statements below.

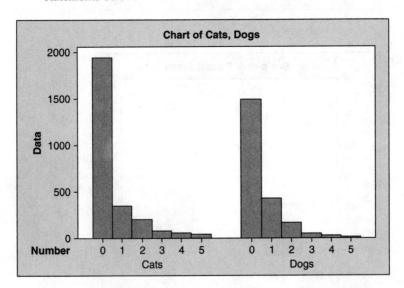

(A) There are more cat owners than dog owners. There are more cats than dogs.
(B) There are more cat owners than dog owners. There are more dogs than cats.
(C) There are more dog owners than cat owners. There are more cats than dogs.
(D) There are more dog owners than cat owners. There are more dogs than cats.
(E) The cat owners are equal to dog owners. There are more cats than dogs.

47. The dataset of water consumption for a small town in gallons per day is listed. What is the effect on the mean and standard deviation if consumption is increased by 50 gallons per day?

{166, 179, 193, 175, 144, 151, 173, 175, 177, 160, 195, 225, 240, 144, 162, 145, 177, 163, 149, 188}

(A) The mean and standard deviation remain the same. *avg distance remain same*

(B) The mean remains the same. The standard deviation increases by $\sqrt{50}$.

(C) The mean increases by 50. The standard deviation remains the same. ✓

(D) The mean and standard deviation increase by 50.

(E) The mean increases by 50. The standard deviation increases by $\sqrt{50}$.

48. A random sample of households and the number of cars per household are shown in the bar chart. What is the **best** estimate of the sample mean?

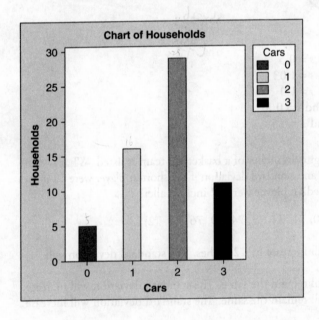

(A) $\bar{X} \approx 1$

(B) $\bar{X} \approx 2$

(C) $\bar{X} \approx 2.5$

(D) $\bar{X} \approx 3$

(E) $\bar{X} \approx 27$

$$\bar{X} = \frac{5(0)+16(1)+2(29)+16(3)}{5+16+29+11} = \frac{?}{8}$$

49. The pie chart shows the number of respondents to a survey that asked how many trips they made to the supermarket in the last week. By reconstructing the frequency distribution, find the mean and standard deviation of the number of trips made to the supermarket.

Respondents

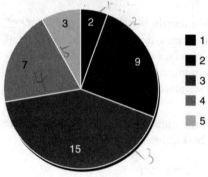

- ■ 1
- ■ 2
- ■ 3
- ■ 4
- ■ 5

(A) $\bar{X} = 3.2$ and $s = 0.5$
(B) $\bar{X} = 3.0$ and $s = 1.03$
(C) $\bar{X} = 3.0$ and $s = 1.0$
(D) $\bar{X} = 108.0$ and $s = 0.3$
(E) $\bar{X} = 21.6$ and $s = 0.5$

50. The dataset of heights in inches of a basketball team is listed. What is the effect on the mean and standard deviation if the shortest player were 12 inches shorter and the median player were 12 inches taller?

{68, 77, 79, 69, 70, 71, 72, 73, 74, 69, 76, 74, 72}

(A) The mean will increase by 12 inches. The standard deviation will increase.

(B) The mean will remain the same. The standard deviation will increase.

(C) The mean will remain the same. The standard deviation will increase by 12 inches.

(D) The mean will remain the same. The standard deviation will remain the same.

(E) The mean and standard deviation will each increase by 12 inches.

51. Choose the five-number summary for the dataset of wingspans in inches of 20 butterflies.

    {3.2, 3.1, 2.9, 4.2, 3.7, 3.5, 4.0, 3.1, 2.9, 3.2, 3.6, 4.1, 3.7, 3.9, 4.1, 3.0, 3.2, 3.8, 3.9, 3.3}

    (A) Min: 2.9; Q1: 3.15; Med: 3.55; Q3: 3.85; Max: 4.2
    (B) Min: 2.9; Q1: 3.1; Med: 3.65; Q3: 3.9; Max: 4.2
    (C) Min: 2.9; Q1: 3.15; Med: 3.65; Q3: 3.9; Max: 4.2
    (D) Min: 2.9; Q1: 3.15; Med: 3.55; Q3: 3.9; Max: 4.2
    (E) Min: 2.9; Q1: 3.1; Med: 3.55; Q3: 3.85; Max: 4.2

52. The cumulative frequency distribution of airfares is displayed in the ogive. What airfares correspond to the 20th percentile, the 50th percentile, and the 80th percentile?

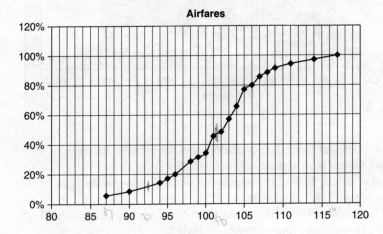

Airfares

    (A) $P_{20} = 95$; $P_{50} = 102$; $P_{80} = 106$
    (B) $P_{20} = 96$; $P_{50} = 103$; $P_{80} = 106$
    (C) $P_{20} = 96$; $P_{50} = 102$; $P_{80} = 106$
    (D) $P_{20} = 96$; $P_{50} = 102$; $P_{80} = 100$
    (E) $P_{20} = 96$; $P_{50} = 102$; $P_{80} = 107$

53. The histogram represents the heights of males in the United States between the ages 20–29. The mean is 69.6 inches and the standard deviation is 2.6 inches. Since the heights are approximately normal, find the height that represents the 20th percentile. How should this be interpreted?

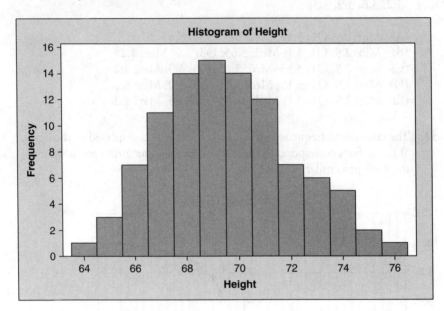

(A) 60 inches. 20% of the population of males between the ages 20–29 are shorter than approximately 60 inches.

(B) 68.6 inches. 20% of the population of males between the ages 20–29 are shorter than approximately 68 inches.

(C) 67.4 inches. 20% of the population of males between the ages 20–29 are shorter than approximately 68 inches.

(D) 67.4 inches. 20 males between the ages 20–29 are shorter than approximately 67 inches.

(E) 60 inches. 20 males between the ages 20–29 are shorter than approximately 60 inches tall.

20th = -.84 (2.6) + 69.6
     = 67.4

54. The histogram represents the heights of males in the United States b...
the ages 20–29. The mean is 69.6 inches and the standard deviation is...
2.6 inches (approximately normal). Three adult males are selected from the
group. Their heights are 75 inches, 63 inches, and 79 inches. Determine
which heights, if any, are unusual.

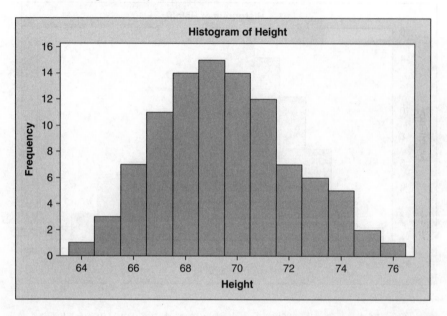

(A) The 79-inch height is the only one that is unusual because it is
more than <u>three standard deviations</u> above the mean. The z-scores,
respectively, are: 2.08, –2.54, and 3.62.

(B) The 79-inch and 63-inch heights are the only ones that are unusual
because they are more than two standard deviations from the mean.
The z-scores, respectively, are: 2.00, –2.54, and 3.62.

(C) All heights are unusual since they are more than two standard
deviations from the mean. The z-scores, respectively, are: –2.08,
–2.54, and 3.62.

(D) All heights are unusual since they are more than two standard
deviations from the mean. The z-scores, respectively, are: 2.08, –2.54,
and 3.62.

(E) All heights are unusual since they are more than two standard
deviations from the mean. The z-scores, respectively, are: 2.08, 2.54,
and 3.62.

phillipa seo

presents the heights of males in the United States between
he mean is 69.6 inches and the standard deviation is
ximately normal). What percentile is a height of
should this be interpreted?

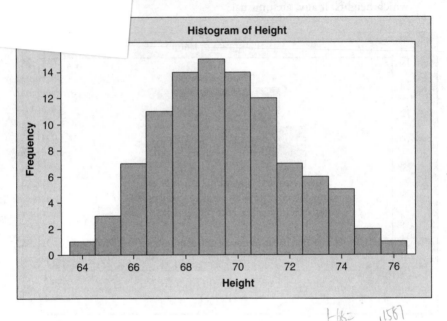

Histogram of Height

$\vdash 66 = .1587$

(A) 16th percentile. 15% of males between the ages 20–29 are shorter
than 67 inches.
(B) 16th percentile. 16% of males between the ages 20–29 are shorter
than 67 inches.
(C) 16th percentile. 16 males between the ages 20–29 are shorter than
67 inches.
(D) 15th percentile. 15 males between the ages 20–29 are shorter than
67 inches.
(E) 16th percentile. 15 males between the ages 20–29 are shorter than
67 inches.

56. Find the mean, median, and mode, if any, for the dataset.

{9, 7, 8, 6, 9, 12, 11, 5, 9, 10}

(A) Mean: 8.6; median: 9; mode: 9
(B) Mean: 8.5; median: 9; mode: 9
(C) Mean: 8.6; median: 8.5; mode: 9
(D) Mean: 8.5; median: 9.5; mode: 9
(E) Mean: 8.0; median: 9; mode: 9

57. Find the variance, standard deviation, and range.

{9, 7, 8, 6, 9, 12, 11, 5, 9, 10}

(A)   Variance: 5.0; standard deviation: 2.2; range: 7
(B)   Variance: 4.7; standard deviation: 2.2; range: 7
(C)   Variance: 4.7; standard deviation: 2.0; range: 7
(D)   Variance: 5.0; standard deviation: 2.2; range: 6
(E)   Variance: 5.0; standard deviation: 2.0; range: 7

58. The frequency distribution shows the number of magazine subscriptions per household in a random sample ($n = 50$). Find the mean number of subscriptions per household.

Number of Magazines, x	0	1	2	3	4	5	6
Frequency	12	8	17	7	3	1	2

(A)   Mean is 2.0
(B)   Mean is 1.8
(C)   Mean is 1.5
(D)   Mean is 3.0
(E)   Mean is 92

$$\frac{0(12)+1(8)+2(17)+3(7)+4(3)+5(1)+6(2)}{50}$$

59. The frequency distribution shows the number of magazine subscriptions per household in a random sample ($n = 50$). Find the standard deviation for the number of subscriptions per household.

Number of Magazines, x	0	1	2	3	4	5	6
Frequency	12	8	17	7	3	1	2

(A)   Standard deviation is 1.5
(B)   Standard deviation is 2.3
(C)   Standard deviation is 3.0
(D)   Standard deviation is 1.2
(E)   Standard deviation is 2.0

$$s = \sqrt{\frac{\sum x^2}{n}} = s = \sqrt{\sum x^2 - \frac{\sum x^2}{n}}$$

60. The mean rate for cable with Internet from a sample of households was $106.50 per month with a standard deviation of $3.85 per month. Assuming the dataset has a normal distribution, estimate the percent of households with rates from $100 to $115.

(A)   94%
(B)   3.2%
(C)   49%
(D)   99%
(E)   6.4%

−1.688   2.2   68-95-99.7

−1.6   2.2

61. Find the mean and median from the datasets below. Which measure of central tendency would be the appropriate measure to use for each? Why?

    Dataset 1: {7, 7, 2, 8, 5, 10, 9, 1, 10, 3}
    Dataset 2: {6, 2, 6, 5, 3, 8, 5, 7, 15, 5}

    (A) Datasets 1 and 2: mean: 6.2; because there are no outliers.
    (B) Dataset 1: mean: 6.2; because there are no outliers. Dataset 2: median: 5.5; because 15 is an outlier.
    (C) Dataset 1: median: 7; because there are no outliers. Dataset 2: mean: 6.2; because 15 is an outlier.
    (D) Datasets 1 and 2: median: 6.2; because 10 and 15 are outliers.
    (E) Datasets 1 and 2: mean: 6.2; because 10 and 15 are outliers.

62. A score at or above the 90th percentile is how many standard deviations from the mean?   *z-score h28*
    (A) 1.28 standard deviations or more above the mean
    (B) 1.64 standard deviations or more above the mean
    (C) 2.58 standard deviations or more above the mean
    (D) 1.28 standard deviations or more below the mean
    (E) 1.64 standard deviations or more below the mean

63. A score at or above the 70th percentile is how many standard deviations from the mean?
    (A) 0.52 standard deviations or more above the mean
    (B) 1.04 standard deviations or more above the mean
    (C) 1.64 standard deviations or more above the mean
    (D) 0.52 standard deviations or more below the mean
    (E) 1.04 standard deviations or more below the mean

64. Test scores are normally distributed with a mean of 72 and a standard deviation of 4.2. What is the percentile of a score of 65 (round to the nearest whole percentile)?
    (A) 95th percentile
    (B) 90th percentile
    (C) 85th percentile
    (D) 5th percentile
    (E) 1st percentile

    $\dfrac{65-72}{4.2}$ *z-score*

    normalcdf (zscore)

65. What data point is indicated on the distribution?

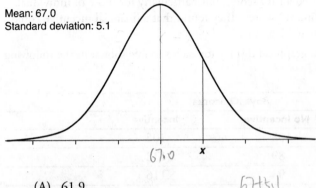

Mean: 67.0
Standard deviation: 5.1

67.0   x

67+5.1

(A) 61.9
(B) 64.5
(C) 67.0
(D) 72.1
(E) 77.2

66. The wage gap between men and women's earnings was 23 cents on the dollar in 2009, down from a low of 22.2 cents in 2007. Some regions in the United States are more progressive and pay is more equitable between genders. "According to WebMD, male specialists earn a median salary of $225,000 a year. Female specialists, on the other hand, take home a median annual salary of $160,000—a difference of $65,000 a year. This difference in pay means that in the course of a 35-year career, female doctors lose a total of $2.3 million on average, noted Forbes."

(a) What might explain the reporting of median salaries rather than average salaries?   mean   extreme outliers

(b) A sample of eight specialist salaries for male and female doctors was collected in one region of the United States. A summary is shown. On average, how does this sample compare with findings by WebMD? How does it compare to the wage gap reported for 2009 of 23 cents on the dollar?   women's lower than men pretty accurate

increase   men: 148+171
                 ———————— = 159.5
                      2   women:
                                          143+126
                                          ——————
                                            2 =

                                          134.5

                                          159.5-134.5
                                          —————————
                                             159.5

                                          11 x
                                          ——
                                          100

                                          x=16

| | Men: Salaries in $1,000 | | | | Women: Salaries in $1,000 | | | |
Category	Mean	Median	Min	Max	Mean	Median	Min	Max
Anesthesiologists	267.9	247	126	395	189.3	194	101	296
Obstetricians and Gynecologists	158.8	148	107	244	143.7	143	78	190
Pediatricians	141.4	146	85	193	112.8	112	81	163
Psychiatrists	157.4	171	87	216	114.1	126	60	160
Surgeons	196	200	124	250	149.1	164	95	192
Podiatrists	97.8	89	50	148	74.3	74	39	116

67. Researchers from the University of Pennsylvania, Philadelphia, found that incentives increased all IQ scores, but particularly for those of individuals with lower baseline IQ scores. They found that a financial incentive could raise IQ as much as 15 points.

(a) Construct a graphical display for each variable shown in the following dataset.

Boys IQ Scores	
No Incentive	Incentive
108	105
89	132
93	116
94	91
123	81
115	83
96	90
73	85
85	131
103	136

(b) Analyze the data for outliers. No

(c) Report an appropriate measure of central tendency for both groups.

(d) What is the IQ score for the 75th percentile for both samples?

68. A basketball team has the following heights and weights.

Heights	Weights
72	180
74	168
68	225
76	201
74	189
69	192
72	197
79	162
70	174
69	171
77	185
73	210

(a) Construct a graphical display for each variable shown in the dataset.
(b) Analyze the data for outliers. *NO*
(c) Report an appropriate measure of central tendency for both weight
    and height. *72.75 , mod 72.5   2. 187.83   med 187   only means*
                                                          (no skew)
(d) What is the height for the 25th percentile?
    69.5   *172.5*

69. Water consumption in gallons per day from a small town is listed below.

Water Consumed
167
180
192
173
145
151
174
175
178
160
195
224
244
146
162
146
177
163
149
188

(a) Construct a frequency distribution table using only five classes.
(b) Construct a graphical display.
(c) Report an appropriate measure of central tendency. *173.5 med*
(d) Analyze the data for outliers using a box plot.

244

70. The following is a sample of butterfly wingspans measured in millimeters.

Wingspans
32
31
29
46
37
38
40
30
28
33
36
39
37
39
41
29
32
38
39
35
37
33

(a) Construct a stem-and-leaf plot.
(b) Report an appropriate measure of central tendency. 35.41 36.5
(c) Analyze the data for outliers. no
(d) What wingspan is the 75th percentile? 39

CHAPTER 3

Two-Variable Data Analysis

71. Which of the following statements **best** describes the scatter plot?

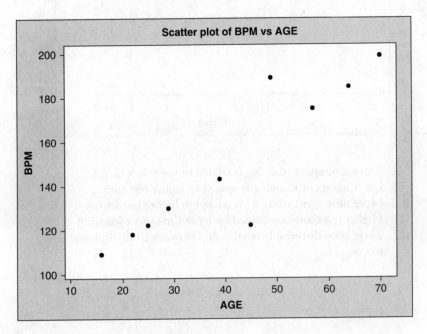

I. A linear model would fit the data well.
II. The bivariate data are positively correlated.
III. The linear relationship is weak (strong)

(A) I only
(B) I and III only
(C) I and II only
(D) II only
(E) III only

72. The scatter plot of study time in hours versus test scores in a statistics class is shown below. What can be said about the relationship between time spent studying and test scores?

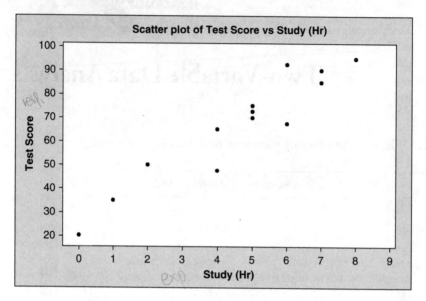

(A) More time spent studying is related to lower test scores.
(B) Less time spent studying is related to higher test scores.
(C) More time spent studying is related to higher test scores.
(D) Higher test scores are caused by more time spent studying.
(E) There is no discernible relationship between study time and test scores.

73. From the data, find the least-squares regression line. What type of correlation is there?

Study (hr)	Test Score
0	20
1	35
2	50
4	47.5
4	65
5	72.5
5	75
6	67.5
6	92.5
7	85
7	90
8	95
5	70

(A) $\hat{y} = 24.765 - 9.051x$. There is negative linear correlation.

(B) $\hat{y} = 24.765 + 9.051x$. There is positive linear correlation.

(C) $\hat{y} = 9.051 + 24.765x$. There is positive linear correlation.

(D) $\hat{y} = 9.051 - 24.765x$. There is negative linear correlation.

(E) $\hat{y} = -24.765 + 9.051x$. There is negative linear correlation.

74. From the data, find the least-squares regression line. What type of correlation is there?

Entertainment Media (hr)	Test Score
0	97.5
1	85
2	82.5
3	75
3	95
5	67.5
5	75
5	82.5
6	57.5
7	65
7	75
10	45

(A) $\hat{y} = 95.494 + 4.508x$. There is positive linear correlation.

(B) $\hat{y} = 4.508 - 95.494x$. There is negative linear correlation.

(C) $\hat{y} = 4.508 + 95.494x$. There is positive linear correlation.

(D) $\hat{y} = 95.494 - 4.508x$. There is negative linear correlation.

(E) $\hat{y} = -95.494 - 4.508x$. There is negative linear correlation.

75. Of the three lines graphs shown below, which is the **best** fitting? Why?

I.   Sum of squares: 5.000

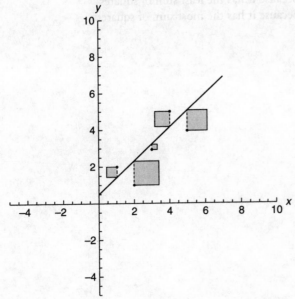

II.   Sum of squares: 4.171

III.

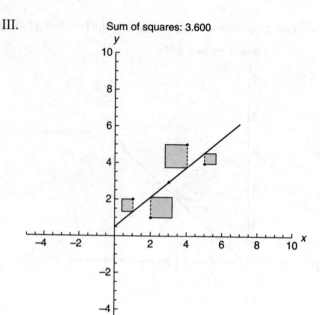

Sum of squares: 3.600

(A) I, because it has the most sum of squares
(B) II, because it has the least sum of squares
(C) II, because it has the most sum of squares
(D) III, because it has the least sum of squares
(E) III, because it has the most sum of squares

76. Of the three regressions, which has the largest coefficient of determination?

I.

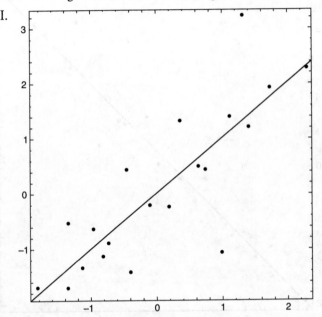

II.

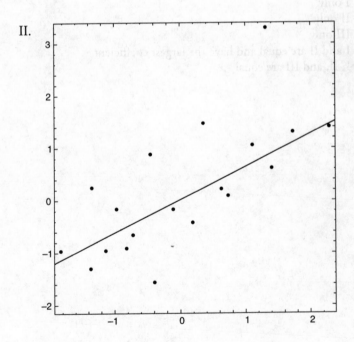

III.

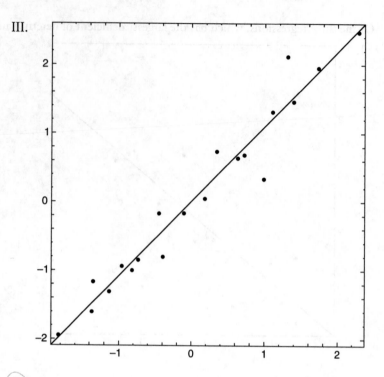

(A) I only
(B) II only
(C) III only
(D) I and II are equal and have the largest coefficient
(E) I, II, and III are equal

77. Which of the plots has a correlation coefficient (r) approximately equal to 0.9?

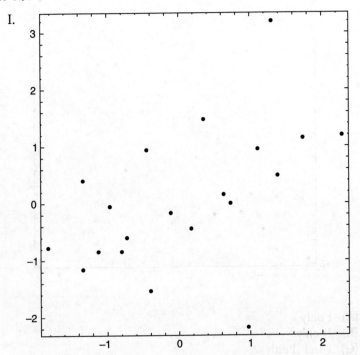

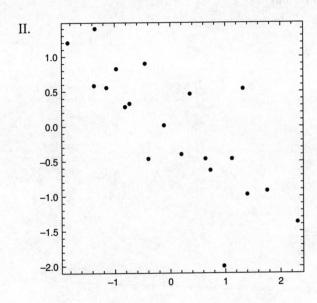

III.

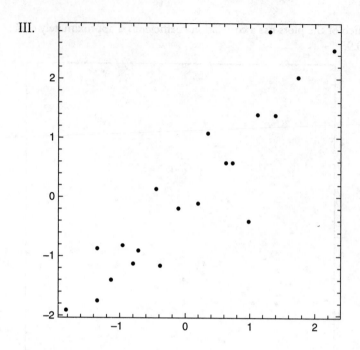

(A) I only
(B) II only
(C) I and II only
(D) III only
(E) II and III only

78. In the residual plot below, is the particular point indicated an overestimate, underestimate, or neither? Why?

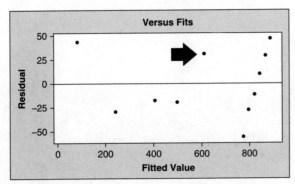

(A) Overestimate, because $y - \hat{y} < 0$
(B) Underestimate, because $y - \hat{y} > 0$ ✓
(C) Neither
(D) Underestimate, because $y - \hat{y} < 0$
(E) Overestimate, because $y - \hat{y} > 0$

79. In the scatter plot, what effect does the indicated point have on the correlation coefficient and the slope of the least-squares regression line?

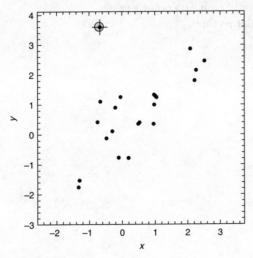

(A) The point is influential; r decreases and the slope increases.
(B) The point is influential; r decreases and the slope decreases.
(C) The point is influential; r decreases and the slope is unaffected.
(D) The point is not influential; therefore r and the slope are unaffected.
(E) The point is influential; therefore r and the slope are unaffected.

80. In the scatter plot, how could the data be transformed in order to do a linear regression?

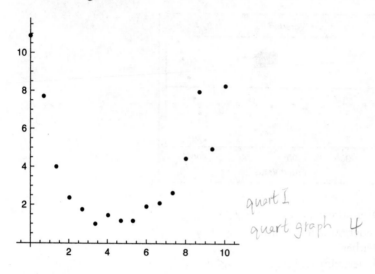

quart I

quart graph   4

(A) By squaring each data point y-value
(B) By taking the natural logarithm of each data point y-value
(C) By raising each data point y-value to the base 10
(D) By taking the square root of each data point y-value
(E) By cubing each data point y-value

81. Using the dataset, which of the following statements **best** describes the correlation?

x secs	y mm
2.5	116
2.75	116
3	130
3.25	125
3.5	129
3.75	137
4	133
4.25	138
4.5	143
4.75	142
5	145
5.25	153
5.5	140
5.75	144
6	142
6.25	140
6.5	132
6.75	125
7	122
7.25	114
7.5	111
7.75	103

I. A linear model would fit the data well.
II. The data point y-value would need to be transformed to achieve linearity. ✓
III. The bivariate data are negatively correlated.

(A) I only
(B) I and III only
(C) I and II only
(D) II only
(E) III only

82. Using the dataset, what can be said about the relationship between earnings and dividends?

Earnings	Dividends
2.78	2.16
1.41	0.88
2.74	1.04
0.92	1.1
2.44	0.96
3.5	2.18
3.68	1.54
1.97	1.39
1.95	1.41
2.07	2.88

(A) More earnings are related to more dividends.
(B) There is no discernible relationship between earnings and dividends.
(C) More earnings are related to less dividends.
(D) More dividends are caused by more earnings.
(E) Fewer earnings are related to more dividends.

83. From the data, find the least-squares regression line. What type of correlation is there?

Calories	Sodium
186	495
181	477
176	425
149	322
184	482
190	587
158	370
139	322
175	479
148	375

(A) $\hat{y} = -299.5 + 4.347x$. There is negative linear correlation.
(B) $\hat{y} = -299.5 + 4.347x$. There is positive linear correlation.
(C) $\hat{y} = 299.5 - 4.347x$. There is positive linear correlation.
(D) $\hat{y} = 299.5 - 4.347x$. There is negative linear correlation.
(E) $\hat{y} = -299.5 + 4.347x$. There is no discernible relationship.

84. From the regression output below, which type of variation does R-squared (R-Sq) pertain to?

Regression Analysis: Sodium versus Calories

The regression equation is: Sodium = −299 + 4.35 Calories

Predictor	Coef	SE Coef	T	P
Constant	−299.5	100.3	−2.99	0.017
Calories	4.3469	0.5917	7.35	0.000

S = 32.6453   R-Sq = 87.1%   R-Sq(adj) = 85.5%

(A) Total variation
(B) Unexplained variation
(C) Explained variation
(D) Least squares variation
(E) Error variation

85. From the regression output below, what percentage of variation is unexplained?

Regression Analysis: Sodium versus Calories

The regression equation is: Sodium = −299 + 4.35 Calories

Predictor	Coef	SE Coef	T	P
Constant	−299.5	100.3	−2.99	0.017
Calories	4.3469	0.5917	7.35	0.000

S = 32.6453   R-Sq = 87.1%   R-Sq(adj) = 85.5%

1−.671

(A) 87.1%
(B) 85.5%
(C) 32.6%
(D) 12.9%
(E) All the variation is explained

86. Find the slope from the least-squares regression line for the data below of final grade as a percent and days absent. What does it tell us about the relationship?

Final	Absences
81	1
90	0
86	2
76	3
51	6
75	4
44	7
81	2
94	0
93	1

(A) Slope is 94.695. For each day absent, the student's final grade increases by 0.947%.

(B) Slope is –94.695. For each day absent, the student's final grade decreases by 0.947%.

(C) Slope is 6.767. For each day absent, the student's final grade increases by 6.767%.

(D) Slope is 6.767. For each day absent, the student's final grade decreases by 6.767%.

(E) Slope is –6.767. For each day absent, the student's final grade decreases by 6.767%.

87. Find the intercept from the least-squares regression line for the data below of final grade as a percent and days absent. What does it tell us about the relationship?

Final	Absences
81	1
90	0
86	2
76	3
51	6
75	4
44	7
81	2
94	0
93	1

(A) Intercept is 94.965. For no absences, the student's predicted grade is 94.965%.

(B) Intercept is 94.695. For no absences, the student's predicted grade is 94.695%.

(C) Intercept is 6.767. For no absences, the student's predicted grade is 67.67%.

(D) Intercept is –6.767. For no absences, the student's predicted grade is 67.67%.

(E) Intercept is 6.676. For no absences, the student's predicted grade is 66.76%.

88. From the regression output and the scatter plot, find r. How would you classify the strength of this relationship?

Regression Analysis: Final versus Absences

The regression equation is: Final = 94.7 – 6.77 Absences

Predictor	Coef	SE Coef	T	P
Constant	94.695	2.376	39.85	0.000
Absences	–6.7672	0.6860	–9.87	0.000

$S = 4.96563$ R-Sq = 92.4%   R-Sq(adj) = 91.5%

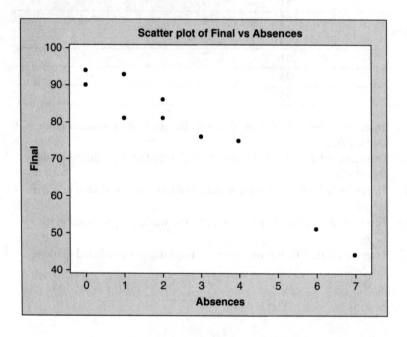

(A) $r = 0.924$. There is a strong positive linear relationship.
(B) $r = -0.924$. There is a strong negative linear relationship.
(C) $r = 0.961$. There is a strong positive linear relationship.
(D) $r = -0.961$. There is a strong negative linear relationship.
(E) $r = 0.924$. There is a strong negative linear relationship.

$$r = -\sqrt{r^2} = -\sqrt{.924} = -.961$$

89. Compare the two scatter plots. Which of the following statements are true about r?

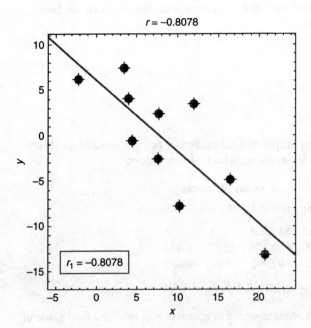

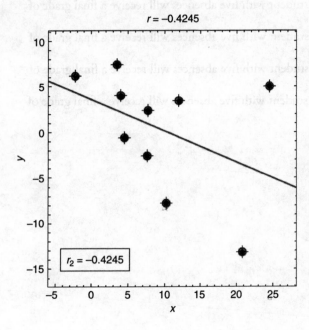

I. r is not resistant to outliers. ✓ single outlier un effect r_1

II. $|r_1| > |r_2|$.

III. r cannot be used to find the percentage of variation that has been explained.

(A) I only
(B) I and III only
(C) I and II only
(D) II only
(E) III only

90. From the regression output of final grade as a percent versus days absent, make a prediction for a student who is absent 5 days.

Regression Analysis: Final versus Absences

The regression equation is: Final = 94.7 − 6.77 Absences

Predictor	Coef	SE Coef	T	P
Constant	94.695	2.376	39.85	0.000
Absences	−6.7672	0.6860	−9.87	0.000

S = 4.96563   R-Sq = 92.4%   R-Sq(adj) = 91.5%

(A) $\hat{y} = 46.673$. A student with five absences will receive a final grade of 46.673%.

(B) $\hat{y} = 68.05$. A student with five absences will receive a final grade of 68.05%.

(C) $\hat{y} = 63.55$. A student with five absences will receive a final grade of 63.55%.

(D) $\hat{y} = 60.85$. A student with five absences will receive a final grade of 60.85%.

(E) $\hat{y} = 60.58$. A student with five absences will receive a final grade of 60.58%.

91. From the dataset below, make a prediction when $x = 19$.

x	y
18.1	31.4
20.0	33.0
20.8	27.4
21.5	25.4
22.1	26.8
22.5	27.8
22.8	23.4
24.0	24.6
25.3	16.8
27.3	18.1

(A) $\hat{y} = 32$
(B) $\hat{y} = 31.5$
(C) $\hat{y} = 35.1$
(D) $\hat{y} = 3.15$
(E) $y = 351$

92. Compare the two regression outputs. Which regression equation explains more of the variation in y?

1. Regression Analysis: Y_1 versus X

The regression equation is: $Y\_1 = 64.8 - 1.75\,X$

Predictor	Coef	SE Coef	T	P
Constant	64.833	6.676	9.71	0.000
X	−1.7542	0.2957	−5.93	0.000

$S = 2.34379$ R-Sq = 81.5%   R-Sq(adj) = 79.2%

2. Regression Analysis: Y_2 versus X

The regression equation is: $Y\_2 = 65.1 - 1.77 X$

Predictor	Coef	SE Coef	T	P
Constant	65.124	6.912	9.42	0.000
X	−1.7697	0.3061	−5.78	0.000

$S = 2.42669$ R-Sq = 80.7%   R-Sq(adj) = 78.3%

(A) Regression 1, since R-Sq, unexplained variation, is the smaller
(B) Regression 2, since R-Sq, total variation, is the smaller
(C) Regression 1, since R-Sq, explained variation, is the greater
(D) Regression 2, since R-Sq, explained variation, is the greater
(E) Regression 1, since R-Sq, total variation, is the greater

93. Find and interpret r for the dataset.

x	y
18.1	31.4
20.0	33.0
20.8	27.4
21.5	25.4
22.1	26.8
22.5	27.8
22.8	23.4
24.0	24.6
25.3	16.8
27.3	18.1

(A) $r = 0.0903$. There is a weak positive linear correlation.
(B) $r = -0.0903$. There is a weak negative linear correlation.
(C) $r = 0.903$. There is a strong positive linear correlation.
(D) $r = -0.903$. There is a strong negative linear correlation.
(E) $r = 0.0903$. There is no linear correlation.

94. Find and interpret the slope for the dataset.

x	y
18.1	31.4
20.0	33.0
20.8	27.4
21.5	25.4
22.1	26.8
22.5	27.8
22.8	23.4
24.0	24.6
25.3	16.8
27.3	18.1

(A) Slope is −1.574. For each unit of x, y decreases by 1.574.
(B) Slope is 1.574. For each unit of x, y decreases by 1.574.
(C) Slope is −1.745. For each unit of x, y decreases by 1.745.
(D) Slope is 1.754. For each unit of x, y decreases by 1.754.
(E) Slope is −1.754. For each unit of x, y decreases by 1.754.

95. From the time series data, find r. What does r tell us about the regression?

Year	Exports ($Mn)
1995	37.1
1996	30.5
1997	29.4
1998	27.4
1999	26.5
2000	29.4
2001	27.1
2002	29.0
2003	29.3
2004	43.9
2005	36.7
2006	43.1

(A) $r = 0.238$. There is a very weak linear relationship.
(B) $r = -0.238$. There is a very weak linear relationship.
(C) $r = 0.488$. There is a very weak positive linear relationship.
(D) $r = -0.488$. There is a very weak negative linear relationship.
(E) $r = -0.488$. There is no discernible relationship.

96. What does the scatter plot of the residuals versus the fitted value $\hat{y}$ tell us about the fit of the data to the model?

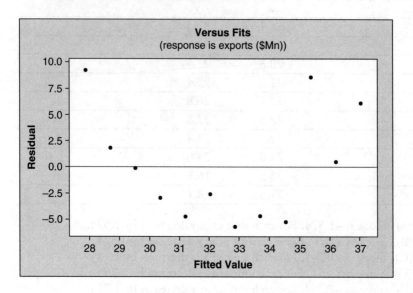

I. A line would fit the data after a transformation.
II. The linear relationship is weak.
III. The bivariate data are negatively correlated. (not really some are +)
(A) I only
(B) I and III only
(C) I and II only
(D) II only
(E) III only

97. What does the scatter plot of the residuals versus the fitted value $\hat{y}$ tell us about the fit of the data to the model?

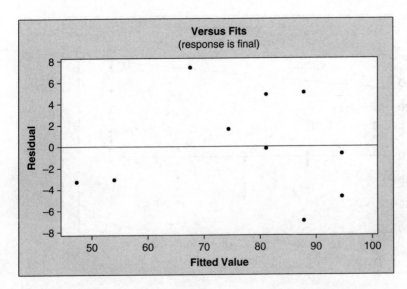

I. A line would fit the data.
II. The linear relationship is strong.
III. The bivariate data are positively correlated. *again like 95b we cannot make the assumption*

(A) I only
(B) I and III only
(C) I and II only
(D) II only
(E) III only

98. What does the scatter plot of the residuals versus the fitted value $\hat{y}$ tell us about the fit of the data to the model?

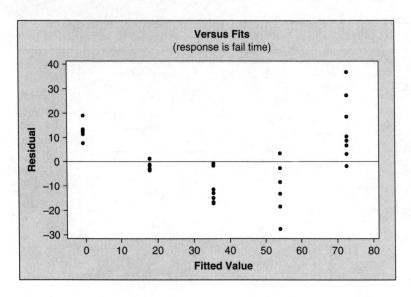

I.   A line would fit the data after a transformation.
II. The linear relationship is weak.
III. The bivariate data are negatively correlated. *There are some*
                                                    datas are positive

(A) I only
(B) I and II only
(C) I and III only
(D) II only
(E) III only

99. From the regression output, find the sum of the squares of the residuals. What type of variation does this represent?

Regression Analysis: Final versus Absences

The regression equation is: Final = 94.7 − 6.77 Absences

Predictor	Coef	SE Coef	T	P
Constant	94.695	2.376	39.85	0.000
Absences	−6.7672	0.6860	−9.87	0.000

$S = 4.96563$ R-Sq $= 92.4\%$ R-Sq(adj) $= 91.5\%$

Analysis of Variance

Source	DF	SS	MS	F	P
Regression	1	2399.6	2399.6	97.32	0.000
Residual Error	8	197.3	24.7		
Total	9	2596.9			

(A) SSE = 197.3; explained variation
(B) SSE = 197.3; total variation
(C) SSE = 197.3; constant variation
(D) SSE = 197.3; standard variation
(E) SSE = 197.3; unexplained variation

100. Compare the two datasets below. Which statements are true?

Absences	Final	Absences	Midterm
1	81	1	75
0	90	0	80
2	86	2	91
3	76	3	80
6	51	6	62
4	75	4	90
7	44	7	60
2	81	2	82
0	94	0	88
1	93	1	96

    I. Absences are a better predictor of the final grade.
   II. The sum of the residuals is the same for both.
  III. The unexplained variation in the final is much greater than in the midterm.

(A) I only
(B) I and III only
(C) I and II only
(D) II only
(E) III only

101. From the regression output below, calculate the coefficient of determination. What does it tell you about the relationship?

Analysis of Variance

Source	DF	SS	MS	F	P
Regression	1	2.8185	2.8185	36.02	0.000
Residual Error	14	1.0955	0.0782		
Total	15	3.9140			

$r^2 = \dfrac{SST - SSE}{SST}$ $\dfrac{2.8185}{3.9140}$

(A) $r^2 = 38.9\%$; that 38.9% of the variation of the residual error is explained by the regression

(B) $r^2 = 44.0\%$; that 44.0% of the variation of the difference of the regression and the residual error is explained by the total

(C) $r^2 = 28.0\%$; that 28.0% of the variation of the response variable is explained by the residuals

(D) $r^2 = 56.3\%$; that 56.3% of the variation of the explanatory variable is explained by the response variable

(E) $r^2 = 72.0\%$; that 72.0% of the variation of the response variable is explained by the explanatory variable

102. From the regression output below, calculate the percentage of unexplained variation in the dependent (response) variable.

Analysis of Variance

Source	DF	SS	MS	F	P
Regression	1	76.4	76.4	0.28	0.606
Residual Error	15	4120.5	274.7		
Total	16	4196.9			

(A) 1.8%
(B) 1.9%
(C) 96.4%
(D) 98.1%
(E) 98.2%

$unexplained = 100\% - r^2\%$

$r^2 = \dfrac{SST - SSE}{SST}$

$r^2 = \dfrac{76.4}{4196.9}$

$r^2 = 1.8\%$

$100\% - 1.8\%$
98.2%

103. Use the dataset to find the correlation coefficient. What would be the **best** explanation of the strength of this relationship?

Expected GPA	Study Time
2.8	8.75
3.4	5.50
2.5	7.00
3.1	7.00
3.0	5.75
2.8	7.50
2.5	5.75
2.8	7.75
2.5	9.50
3.1	4.75
3.0	4.50
3.3	5.75
3.6	3.00
3.3	4.00
2.8	7.00
2.9	7.50
3.1	6.50
3.1	4.00
3.6	4.00

Study time in hours per week

(A) Correlation coefficient is −0.730. That higher grades are earned by more efficient study techniques.
(B) Correlation coefficient is 0.730. That students expecting higher grades study more hours per week.
(C) Correlation coefficient is −0.730. That less time spent studying results in better grades.
(D) Correlation coefficient is −0.730. That a higher expected GPA predicts a lower study time.
(E) Correlation coefficient is 0.730. That more time spent studying results in better grades.

Interpret wrong

104. Use the regression analysis outputs below to compare the two regressions. Which statements are true?

Regression Analysis: Grade to Date versus Points Attempted

The regression equation is: Grade to Date = –0.0026 + 0.000373 Points Attempted

Predictor	Coef	SE Coef	T	P
Constant	–0.00257	0.02444	–0.11	0.917
Points Attempted	0.00037282	0.00002603	14.32	0.000

S = 0.0600739   R-Sq = 90.7%   R-sq(adj) = 90.3%

Regression Analysis: Grade to Date versus Online Activity

The regression equation is: Grade to Date = 0.256 + 0.000373 Online Activity

Predictor	Coef	SE Coef	T	P
Constant	0.25650	0.03956	6.48	0.000
Online Activity	0.0003727	0.0001472	2.53	0.019

S = 0.172573   R-Sq = 23.4%   R-sq(adj) = 19.7%

   I. Points attempted is a better predictor of the grade to date.
  II. The sum of the residuals is the same for both. *always* 0
 III. The unexplained variation in the grade to date is much greater when it is regressed on points attempted than on online activity.

(A) I only
(B) II and III only
(C) I and II only
(D) II only
(E) III only

105. Using the dataset, find the least-squares regression line. Make a prediction when $x = 900$.

Points Attempted	Grade to Date
1124	45.49%
250	11.25%
270	10.80%
873	40.53%
1465	52.45%
1160	45.44%
120	2.65%
1160	43.83%
20	0.91%
770	39.14%
692	16.78%
770	32.82%
1051	24.49%
1593	51.23%
140	7.73%
380	8.70%
1559	59.33%
1708	68.29%
887	34.96%
551	11.47%
602	21.52%
613	21.13%
790	34.66%

(A) $\hat{y} = -0.0026 + 0.000373x$. $x = 900$, $\hat{y} = 3.331$.

(B) $\hat{y} = 0.0026 + 0.000373x$. $x = 900$, $\hat{y} = .3383\%$.

(C) $\hat{y} = -0.0026 + 0.000373x$. $x = 900$, $\hat{y} = .3331\%$.

(D) $\hat{y} = 0.0026 + 0.000373x$. $x = 900$, $\hat{y} = 33.83\%$.

(E) $\hat{y} = -0.0026 + 0.000373x$. $x = 900$, $\hat{y} = 33.31\%$.

106. Find the slope from the least-squares regression line for the data below. What does it tell us about the relationship?

Points Attempted	Grade to Date
1124	45.49%
250	11.25%
270	10.80%
873	40.53%
1465	52.45%
1160	45.44%
120	2.65%
1160	43.83%
20	0.91%
770	39.14%
692	16.78%
770	32.82%
1051	24.49%
1593	51.23%
140	7.73%
380	8.70%
1559	59.33%
1708	68.29%
887	34.96%
551	11.47%
602	21.52%
613	21.13%
790	34.66%

(A) Slope is –0.000373. For every point attempted, the grade to date increases by 0.0373%.

(B) Slope is 0.000373. For every point attempted, the grade to date increases by 0.0373%.

(C) Slope is –0.0026. For every point attempted, the grade to date decreases by 0.26%.

(D) Slope is 0.0026. For every point attempted, the grade to date increases by 0.26%.

(E) Slope is 0.000373. For every percentage point attempted, the grade to date increases by 0.0373%.

107. Find the intercept from the least-squares regression line for the data below. What does it tell us about the relationship?

Points Attempted	Grade to Date
1124	45.49%
250	11.25%
270	10.80%
873	40.53%
1465	52.45%
1160	45.44%
120	2.65%
1160	43.83%
20	0.91%
770	39.14%
692	16.78%
770	32.82%
1051	24.49%
1593	51.23%
140	7.73%
380	8.70%
1559	59.33%
1708	68.29%
887	34.96%
551	11.47%
602	21.52%
613	21.13%
790	34.66%

(A) The intercept is –0.0026. When $x = 0$, the model predicts a grade to date of –0.26%.

(B) The intercept is 0.0026. When $x = 0$, the model predicts a grade to date of 0.26%.

(C) The intercept is –0.0026. When $x = 0$, the model predicts a grade to date of –0.0026%.

(D) The intercept is 0.000373. When $x = 0$, the model predicts a grade to date of –0.26%.

(E) The intercept is 0.000373. When $x = 0$, the model predicts a grade to date of 0.0373%.

108. From the regression output and the scatter plot, find r. How would you classify the strength of this relationship?

The regression equation is: Grade to Date = –0.0026 + 0.000373 Points Attempted

Predictor	Coef	SE Coef	T	P
Constant	–0.00257	0.02444	–0.11	0.917
Points Attempted	0.00037282	0.00002603	14.32	0.000

$S = 0.0600739$ R-Sq = 90.7%   R-Sq(adj) = 90.3%

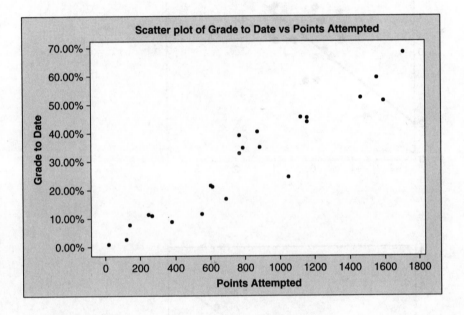

(A) $r = 0.952$. A strong positive linear relationship.
(B) $r = 0.952$. A moderate positive linear relationship.
(C) $r = 9.52$. A moderate positive linear relationship.
(D) $r = 9.52$. A weak positive linear relationship.
(E) $r = -0.952$. A strong negative linear relationship.

109. Compare the two scatter plots. Which of the following statements are true about the regressions?

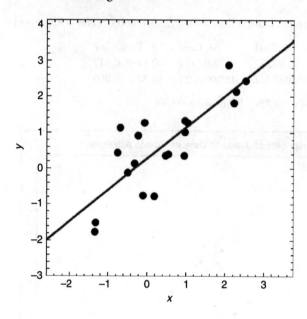

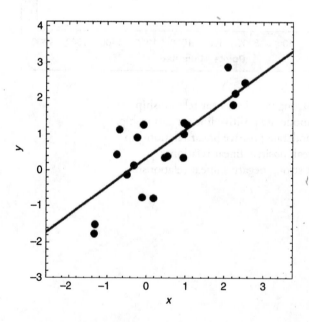

I. Regression 1 explains more variation in the response ✓
II. $r_1^2 > r_2^2$ ✓
III. Can't tell which regression explains more

(A) I only
(B) I and III only
(C) I and II only
(D) II only
(E) III only

110. From the regression output for the data provided in **Questions 107, 108, and 109**, make a prediction when Points Attempted = 3000. How would you classify this prediction? What, if any, limitations would you apply?

The regression equation is: Grade to Date = − 0.0026 + 0.000373 Points Attempted

Predictor	Coef	SE Coef	T	P
Constant	−0.00257	0.02444	−0.11	0.917
Points Attempted	0.00037282	0.00002603	14.32	0.000

S = 0.0600739   R-Sq = 90.7%   R-Sq(adj) = 90.3%

(A) $\hat{y} = 111.64\%$. This is an extrapolation. Grades cannot be greater than 100%. This model would be limited to predictions less than or equal to the maximum possible total points.
(B) $\hat{y} = 11.164\%$. This is an interpolation. This model would be limited to predictions less than or equal to total points.
(C) $\hat{y} = 1116.4\%$. This is an extreme extrapolation. Grades cannot be greater than 100%. This model would be limited to predictions less than or equal to total points.
(D) $\hat{y} = 11164\%$. This is an extreme extrapolation. Grades cannot be greater than 100%. This model would be limited to predictions less than or equal to total points.
(E) $\hat{y} = 111.64\%$. Grades can be greater than 100%. This model is unlimited and can be used to predict any grade.

111. From the dataset below, using the year as the *x*-value for each row, make a prediction when Year = 2075. Would this be appropriate? Why or why not?

100 m Speed Demons		Date	Time
Donald Lippincott	USA	Sat 06 Jul 1912	10.6
Charles Paddock	USA	Sat 23 Apr 1921	10.4
Percy Williams	Can	Sat 09 Aug 1930	10.3
Jesse Owens	USA	Sat 20 Jun 1936	10.2
Willie Williams	USA	Fri 03 Aug 1956	10.1
Armin Hary	Ger	Tue 21 Jun 1960	10.0
Jim Hines	USA	Mon 14 Oct 1968	9.95
Calvin Smith	USA	Sun 03 Jul 1983	9.93
Carl Lewis	USA	Sat 24 Sep 1988	9.92
Leroy Burrell	USA	Fri 14 Jun 1991	9.90
Carl Lewis	USA	Sun 25 Aug 1991	9.86
Leroy Burrell	USA	Wed 06 Jul 1994	9.85
Donovan Bailey	Can	Sat 27 Jul 1996	9.84
Maurice Greene	USA	Wed 16 Jun 1999	9.79
Tim Montgomery	USA	Sat 14 Sep 2002	9.78

(A) $\hat{y} = 9.19$. This is an inappropriate extrapolation.
(B) $\hat{y} = 9.19$. This is an appropriate extrapolation.
(C) $\hat{y} = 9.19$. This is an appropriate interpolation.
(D) $\hat{y} = 9.91$. This is an inappropriate extrapolation.
(E) $\hat{y} = 9.91$. This is an appropriate extrapolation.

112. What assumptions should you verify for a valid least-squares regression line?

(A) That the sample was randomly selected only
(B) That the variance is constant only
(C) That the sample was randomly selected and that variance is constant
(D) That the residuals are normally distributed only
(E) That the sample was randomly selected, that variance is constant, and that the residuals are normally distributed

113. In the scatter plot of residuals versus the fitted value $\hat{y}$, what can you say about the variance of y?

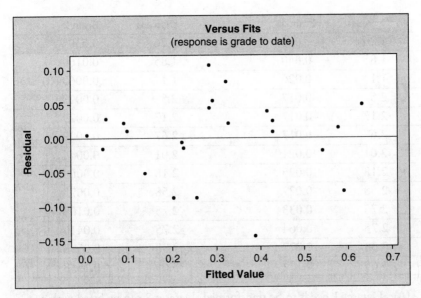

(A) The variance of y increases in the extremes of y.
(B) The variance of y is constant.
(C) The variance of y increases as y increases.
(D) The variance of y increases near the mean of y.
(E) The variance of y decreases near the mean of y.

114. Compare the residuals from the two datasets. Which of the following statements is true?

Distress	Brain 1	Distress	Brain 2
1.26	−0.055	1.26	0.0303
1.85	−0.040	1.85	0.0160
1.1	−0.026	1.1	0.0068
2.5	−0.017	2.5	0.0029
2.17	−0.017	2.17	0.0029
2.67	0.017	2.67	0.0029
2.01	0.021	2.01	0.0044
2.18	0.025	2.18	0.0063
2.58	0.027	2.58	0.0073
2.75	0.033	2.75	0.0109
2.75	0.064	2.75	0.0410
3.33	0.077	3.33	0.0593
3.65	0.124	3.65	0.1538

(A) Dataset 1 needs to be transformed. Dataset 2 can be fitted with a linear regression.
(B) Dataset 1 can be fitted with a linear regression. Dataset 2 needs to be transformed.
(C) Dataset 1 and Dataset 2 can be fitted with a linear regression.
(D) Dataset 1 and Dataset 2 need to be transformed.
(E) There is no indication of a relationship between the variables in either dataset.

115. Given two linear models for the dataset, which is the best? Why?

Correlations: Distress, Brain 1

Pearson correlation of Distress and Brain 1 = 0.878
P-Value = 0.000

I. **Regression Analysis: Brain 1 versus Distress**

The regression equation is: Brain 1 = $- 0.126 + 0.0608$ Distress

Predictor	Coef	SE Coef	T	P
Constant	−0.12608	0.02465	−5.12	0.000
Distress	0.060782	0.009979	6.09	0.000

S = 0.0250896   R-Sq = 77.1%   R-Sq(adj) = 75.1%

Analysis of Variance

Source	DF	SS	MS	F	P
Regression	1	0.023353	0.023353	37.10	0.000
Residual Error	11	0.006924	0.000629		
Total	12	0.030277			

II. **Regression Analysis: Brain 1 versus Distress**

The regression equation is: Brain 1 = 0.0118 Distress

Predictor	Coef	SE Coef	T	P
No Constant				
Distress	0.011807	0.004959	2.38	0.035

S = 0.0441576

Analysis of Variance

Source	DF	SS	MS	F	P
Regression	1	0.011054	0.011054	5.67	0.035
Residual Error	12	0.023399	0.001950		
Total	13	0.034453			

(A) The models are equally good since r is the same.
(B) The models are equally good because the total sum of the squares is equal.
(C) Model II is the best because the total sum of the squares is less than Model I.
(D) Model I is the best because the total sum of the squares is less than Model II.
(E) Neither model explains enough of the variation in the variable Brain 1.

116. A study on grade inflation measured study time in hours per week versus expected grades. A scatter plot for actual data from the mathematics department is shown.

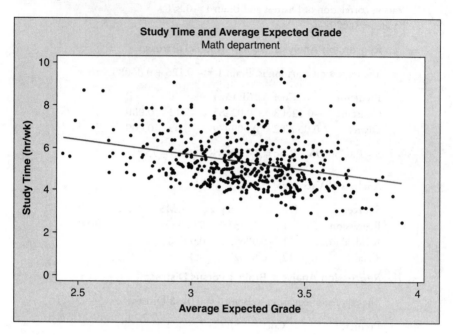

(a) Estimate the a and b values for the linear model shown ($y = a + bx$).

(b) Using your linear model from(A), predict the amount of study time when the average expected grade is 3.25.

(c) The actual linear model from this study is $y' = 10.012 - 1.416x$. Using this model, predict the amount of study time when the expected grade is 3.25.

(d) What is surprising about this model?

117. In an online Algebra 1 course, the following data on points attempted and grade to date were collected one month before the end of the semester.

Points Attempted	Grade to Date
1124	45.49%
250	11.25%
270	10.80%
873	40.53%
1465	52.45%
1160	45.44%
120	2.65%
1160	43.83%
20	0.91%
770	39.14%
692	16.78%
770	32.82%
1051	24.49%
1593	51.23%
140	7.73%
380	8.70%
1559	59.33%
1708	68.29%
887	34.96%
551	11.47%
602	21.52%
613	21.13%
790	34.66%

(a) Find the a and b values for the least-squares regression line $(y = a + bx)$.

(b) Calculate r for the relationship between the grade to date and points attempted. How would you describe the relationship?

(c) Address the assumptions for the validity of a least-squares regression. Does this dataset violate any of those assumptions? If so, how could you overcome that violation?

(d) Do the data in the table above seem to support the teacher's claim that "You know what you do?"

(e) Consider the dataset of the grade to date versus online activity. What might explain why this set contradicts the teacher?

Online Activity	Grade to Date
78	45.49%
54	11.25%
35	10.80%
0	40.53%
90	52.45%
7	45.44%
0	2.65%
205	43.83%
0	0.91%
0	39.14%
55	16.78%
25	32.82%
20	24.49%
94	51.23%
0	7.73%
154	8.70%
192	59.33%
1218	68.29%
15	34.96%
185	11.47%
19	21.52%
78	21.13%
43	34.66%

118. In an Introductory Probability and Statistics course, students did their homework online. They were allowed to resubmit their homework over and over again until the final due date. The dataset below shows the average number of submissions and the average grade for that assignment.

Average Submissions	Average Grade
1.84	73.7%
2.00	90.8%
1.25	89.9%
1.87	95.7%
2.06	88.5%
1.48	90.6%
1.65	76.0%
1.40	97.1%
2.38	90.5%
1.28	94.1%
1.50	90.8%
1.43	87.5%
1.39	94.1%
1.62	91.9%
1.36	97.2%
1.48	76.5%
1.27	98.3%
1.17	96.2%
1.18	88.9%
1.30	96.8%

(a) Analyze the relationship between the two variables. Include a scatter plot.

(b) The instructor thought that students would be motivated to redo an assignment by the incentive of a higher grade. Do these data seem to support that? Why or why not?

(c) Suppose that students are likely to put off doing the assignment until a few hours before the assignment is due. How might that counter the benefit of being able to redo an assignment? Would you consider it a lurking variable?

119. Are reading levels and mathematics grades related?

(a) Analyze the relationship between the two variables in the dataset. Include a scatter plot.

Reading	Math
19	11
17	15
15	9
14	13
13	11
11	9
11	8
9	7
7	6
4	1

(b) What is the model from a least-squares regression line? Use it to predict a math score when the reading score is 12.

(c) How much of the variation in math scores is explained by reading scores?

(d) Check the assumption of constant variance. If the assumption is violated, what might the researcher do?

120. A study on grade inflation published this scatter plot with the least-squares regression line.

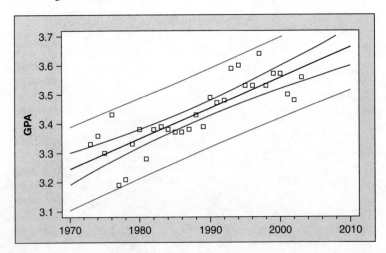

(a) Estimate the linear model shown.
(b) Using your linear model from(A), predict the GPA in the year 2016.
(c) The actual linear model from this study is: $\hat{y} = -16.451 + 0.010x$. Using this model, predict the GPA in the years 2016 and 2036. Are you surprised? Why?
(d) What limitations, if any, would you use to constrain this model?

CHAPTER 4

Design of a Study: Sampling, Surveys, and Experiments

121. A farmer wishes to study the effect of three different fertilizers on crop yields. He takes a rectangular field and divides it into four plots of equal area. Then he randomly assigns the three different fertilizers to one of the four plots. One plot receives no fertilizer. The plots are harvested after a growing period and the yields are measured and compared. Which of the following statements **best** describes the design of the study?

  I. This design has matched pairs.
  II. This design has blocks.
  III. This is a completely randomized design.

  (A) I only
  (B) I and III only
  (C) I and II only
  (D) II only
  (E) III only

122. Which of the following statements is true about completely randomized design?

  I. It replicates treatments to paired groups with similar experimental units.
  II. Treatments are assigned haphazardly to experimental units.
  III. It controls for the effects of lurking variables.

  (A) I only
  (B) I and III only
  (C) I and II only
  (D) II only
  (E) III only

123. Which of the following statements is true about block design?
   I. Treatments are assigned haphazardly to experimental units.
   II. It controls for the effects of confounding variables.
   III. Experimental units are examined before receiving treatments.

   (A) I only
   (B) I and III only
   (C) I and II only
   (D) II only
   (E) III only

124. Which of the following statements is true about matched-pairs design?
   I. It is a type of block design.
   II. It uses the same or similar experimental units with different treatments.
   III. Treatments are assigned haphazardly to experimental units.

   (A) I only
   (B) I and III only
   (C) I and II only
   (D) II only
   (E) III only

125. A woman insists that she can detect whether tea was poured into milk or whether milk was poured into tea. An experiment is devised. In the kitchen, six cups are labeled A, B, C, D, E, and F. Which cups receive tea poured into milk and which receive milk poured into tea is decided by a series of coin flips and recorded. The woman is asked to taste each cup and identify the type of preparation. Which of the following statements best describes the design of the study?
   I. Double-blind experiment
   II. Block design
   III. Completely randomized design

   (A) I only
   (B) I and III only
   (C) I and II only
   (D) II only
   (E) III only

126. A pharmaceutical research company wishes to find the optimal dosage of a drug. The dosages are 250 mg, 500 mg, 750 mg, and 1,000 mg. In Russia, 1,000 experimental subjects are recruited and divided into five groups. The researchers first determine the subjects' height, weight, blood pressure, and gender. Using these metrics, they select five similar individuals and randomly assign each to one of the five groups. Dosages are then randomly assigned to the groups, with one group receiving a placebo. The participants and researchers are not informed of the assignment of dosages. Which of the following statements best describes the design of the study?

   I. Block design
   II. Observational study
   III. Double-blind experiment

   (A) I only
   (B) I and III only
   (C) I and II only
   (D) II only
   (E) III only

127. The purpose of doing an experiment is to:

   (A) Determine cause and effect
   (B) Identify confounding variables
   (C) Identify lurking variables
   (D) Control participants
   (E) Find human error

128. The purpose of blocking in experimental design is to:

   (A) Control for confounding variables
   (B) Control experimenter bias
   (C) Control for lurking variables
   (D) Replicate the experiment
   (E) Organize experimental units

129. The purpose of matched-pairs design is to:

   (A) Control experimenter bias
   (B) Control for lurking variables
   (C) Control for confounding variables
   (D) Replicate the experiment
   (E) Organize experimental units

130. Suppose you wish to study the effects of high-protein kibble on adult dog behavior. What would be the best design to determine between-group effects?

(A) Matched-pairs design
(B) Completely randomized design
(C) Block design
(D) Double-blind design
(E) Randomized design

131. Suppose you wish to study anti-aging effects of an oil extract on women over 50. One hundred participants apply the oil on the right side of their faces for six weeks. Then the researcher examines the right side against the left side for the effects of the oil. Which of the following statements **best** describes the design of the study?

   I. Completely randomized design
  II. Matched-pairs design
 III. Block design

(A) I only
(B) I and III only
(C) I and II only
(D) II only
(E) III only

132. Suppose you want to study the effects of a new organic flea shampoo for dogs. You wish to advertize that it is as effective as the leading brand used by groomers. You have 100 beagles to work with and you have decided to use a completely randomized design. What process would be correct?

(A) Measure the shine of each beagle's coat. Put the 50 shiniest in the group receiving the organic flea shampoo treatment and treat the others with the leading brand.
(B) Select pairs of beagles that have similar to identical coat characteristics. Flip a coin to decide which beagle is treated with the organic flea shampoo.
(C) Randomly assign the beagles to one of two groups. Then randomly assign the two treatments to the groups.
(D) List all the dogs' names in alphabetical order and assign the first 50 to the organic flea shampoo treatment and the others to the leading brand.
(E) Randomly select 50 dogs to receive the organic flea shampoo. The rest receive the leading brand.

133. A new drug for appetite control was tested on 1,000 men and women using a completely randomized design. The researchers expected that the overall effect would be significant weight loss; however, they were disappointed when the group effects were insignificant. When they examined the results, they found that men gained weight while women lost weight on the new drug. What is the **best** explanation for the effects of this experiment?

(A) Experimenter bias was systematic when recording the results.
(B) A placebo effect interfered with the results.
(C) The scale used to measure the weights of men was broken.
(D) Gender is a confounding variable in this study.
(E) Gender is a lurking variable in this study.

134. A new treadmill is compared with an older model. One thousand men and women are blocked by gender and randomly assigned to one of the two machines. Researchers measure each participant's fitness level before the experiment. After a period of time, the researchers measure the fitness levels again. The results are not significantly different. When the participants assigned to the new treadmill are interviewed after the experiment, researchers learn that they did not follow the exercise schedule faithfully because the new treadmill was too complicated to operate. What is the best explanation for the outcomes of the study?

(A) Experimenter bias was systematic when recording the results.
(B) The operation of the new treadmill is a lurking variable in this study.
(C) The device used to measure the fitness levels of men was broken.
(D) A placebo effect interfered with the results.
(E) Gender is a confounding variable in this study.

135. Which of the statements describes good practices when designing experiments?
  I. Control the effects of confounding variables.
  II. Replicate to increase variation from treatments.
  III. Randomize to ensure that an effect is observed.

(A) I only
(B) I and III only
(C) I and II only
(D) II only
(E) III only

136. When the difference between what we expect and what we observe in a study is unusually large and cannot be attributed to chance, the finding is
_____.

    (A) remarkable
    (B) repeatable
    (C) statistically significant
    (D) insignificant
    (E) systematic

137. A new treatment for overactive teenagers called "talk-therapy" is tested on 20 teens. The teens are randomly assigned to two groups. Then one group is randomly assigned the talk-therapy treatment and the other group receives fake talk-therapy. If a teen knows that he or she is getting the real talk-therapy, he or she is less active. When the researchers know they are measuring a teen who received talk-therapy, they systematically lower the ratings. Which type of design would be ideal for controlling for these problems?
    I. Completely randomized design
    II. Double-blind experiment
    III. Matched-pairs design

    (A) I only
    (B) I and III only
    (C) I and II only
    (D) II only
    (E) III only

138. What statements are true about statistical significance?
    I. The difference between what is expected and what is observed is too large to attribute to chance alone.
    II. It determines which experimental design to use.
    III. It is determined numerically with hypothesis testing.

    (A) I only
    (B) I and III only
    (C) I and II only
    (D) II only
    (E) III only

139. Under what conditions would it be necessary to conduct an observational study?
    (A) When there aren't any willing volunteers to participate in the study
    (B) When the treatments aren't effective in a laboratory
    (C) When the subjects can't be randomly selected
    (D) When it is unethical to impose a treatment in order to measure a response on subjects
    (E) When the research is not sufficiently funded

140. What are the advantages of an experiment over an observational study?
    I. Outcomes will naturally change over time.
    II. The researcher has more control of the environment and treatment.
    III. There is more diversity in the sample of subjects.
    (A) I only
    (B) I and III only
    (C) I and II only
    (D) II only
    (E) III only

141. At a charter high school, administrators wish to collect a sample of 50 students. The proportions of the student body represented by each class are: 45% freshmen, 28% sophomores, 16% juniors, and 11% seniors. They decide to randomly sample 23 freshmen, 14 sophomores, eight juniors, and five seniors. Which of the following methods was used to select the probability sample?
    (A) Random
    (B) Systematic
    (C) Stratified
    (D) Cluster
    (E) Simple random

142. The athletic department at a very large high school wishes to form a kickball team of 20 students from the entire student body. They run a lottery to select 20 student ID numbers from a thoroughly mixed bucket of tickets where each student is represented by one ticket. Which of the following methods was used to select the probability sample?
    (A) Random
    (B) Systematic
    (C) Stratified
    (D) Cluster
    (E) Simple random

143. At the Westminster Dog Show, sponsors wish to measure the popularity of the booths in the retail area. Using a list of all attendees, a random number is generated to determine a starting position in the list and then every tenth attendee is placed in the sample. When the end of the list is reached, sampling continues from the top of the list. Sampling stops when the starting position is reached. Which of the following methods was used to select the probability sample?

(A) Random
(B) Systematic
(C) Stratified
(D) Cluster
(E) Simple random

144. A charter school operator in Los Angeles wishes to gather information about student achievement. From the 73 small schools the operator manages, one school is selected by lottery and all students from that school are used in the sample. Which of the following methods was used to select the probability sample?

(A) Random
(B) Systematic
(C) Stratified
(D) Cluster
(E) Simple random

145. The mathematics department wishes to form a sample from each of 10 courses offered. They randomly select three individuals from each course and include them in the 30-person sample. Which of the following methods was used to select the probability sample?

(A) Random
(B) Systematic
(C) Stratified
(D) Cluster
(E) Simple random

146. You would like to determine the characteristics of Facebook users, so you examine the accounts of several Facebook friends and decide that your network of friends is similar to those you observe in other accounts. You use your friends as a sample to determine characteristics of Facebook users. Which of the following methods was used to select the nonrandom sample?

(A) Non-response
(B) Convenience
(C) Quota
(D) Cluster
(E) Self-selected or voluntary response

147. In a statistics course, the instructor wishes to determine whether a new program for statistical analysis would be beneficial to students. He asks for 10 volunteers from the class to use the program in their final project and provide feedback regarding ease of use. Which of the following methods was used to select the nonrandom sample?

(A) Non-response
(B) Convenience
(C) Quota
(D) Cluster
(E) Self-selected or voluntary response

148. A pollster wishes to determine which candidate is most popular among retirees on the issue of health care. Based on party registration proportions, the pollster selects a focus group of 30 retirees. Which of the following methods was used to select the nonrandom sample?

(A) Non-response
(B) Convenience
(C) Quota
(D) Cluster
(E) Self-selected or voluntary response

149. Which of the following **best** describes the goal of sampling?
  I. Produce an average sample.
  II. Choose the average representatives.
  III. Produce a representative sample.

(A) I only
(B) I and III only
(C) I and II only
(D) II only
(E) III only

150. Street parking in your neighborhood is limited. You have amassed 10 parking tickets in one year and believe your neighbors have had a similar experience. You leave 50 postcards on the windshields of parked cars in your neighborhood asking for a response to one question with room for additional comments. You receive 48 responses, and 30 of the responses include elaborate descriptions of street parking issues. Which type of sampling bias **best** describes the situation?

(A) Voluntary response
(B) Wording
(C) Response
(D) Researcher
(E) Non-response

151. Local schools wish to add a bond measure to the ballet. They survey the residents with the following question: "Will you support the bond measure if it will cost the taxpayers $30 million in 30 years?" Three-fourths of respondents reply "No." Which type of sampling bias best describes the situation?

(A) Voluntary response
(B) Wording
(C) Response
(D) Researcher
(E) Non-response

152. A credit card company wishes to sample its customers to find out which features of the rewards program are preferred. Its call center schedules calls between 5:30 and 8:00 P.M. Only 25% of the customers the call center reaches are willing to participate. Which type of sampling bias best describes the situation?

(A) Voluntary response
(B) Wording
(C) Response
(D) Researcher
(E) Non-response

153. A study on sexual activity of students in a college dormitory showed that men reported five casual sexual encounters with women on average while women reported only two over the same time period. Which type of sampling bias **best** describes the situation?

(A) Voluntary response
(B) Wording
(C) Response
(D) Researcher
(E) Non-response

154. Which of the statements is true about experiments?
  I. They impose treatments on subjects.
  II. The researcher doesn't attempt to manipulate a response.
  III. The researcher wishes to determine an effect from a treatment.

(A) I only
(B) I and III only
(C) I and II only
(D) II only
(E) III only

155. Which of the statements is true about observational studies?
  I. They impose treatments on subjects.
  II. The researcher doesn't attempt to manipulate a response.
  III. The researcher wishes to determine an effect from a treatment.

(A) I only
(B) I and III only
(C) I and II only
(D) II only
(E) III only

156. Which of the statements is true about the placebo effect?
  I. It is a response due to believing that a treatment is being administered.
  II. It is an effect from a legitimate treatment.
  III. It is a negative response due to a high dosage.

(A) I only
(B) I and III only
(C) I and II only
(D) II only
(E) III only

157. Which of the statements is true about a double-blind experiment?
   I. Subjects are not aware whether they are receiving viable treatments.
   II. It is a study of adolescent-onset blindness.
   III. Researchers are not aware of the distribution of placebos or viable treatments.

   (A) I only
   (B) I and III only
   (C) I and II only
   (D) II only
   (E) III only

158. In general, which of the following do researchers attempt to achieve when designing experiments?
   I. Attempt to impose statistical control
   II. Attempt to replicate an observational study
   III. Attempt to randomize events

   (A) I only
   (B) I and III only
   (C) I and II only
   (D) II only
   (E) III only

159. What is an appropriate method for performing randomization?

   (A) Appropriately using a table of random digits to select participants or assign treatments
   (B) Haphazardly generating a list of digits to select participants or assign treatments
   (C) Using telephone numbers as random digits to select participants or assign treatments
   (D) Using a complicated mathematical function to select participants or assign treatments
   (E) Rearranging the digits to select participants or assign treatments

160. Which of the statements is true about blocking?
   I. Treatments are organized in the same way between blocks.
   II. They are used to replicate treatments on the same subject.
   III. Treatments are randomized within blocks.

   (A) I only
   (B) I and III only
   (C) I and II only
   (D) II only
   (E) III only

161. If the difference between a treatment and control group is too big to attribute to chance, it is _____.

(A) statistically significant
(B) not significant
(C) important
(D) a maximum
(E) substantial

162. A researcher randomly selected 50 volunteers to participate in a study on weight and energy consumption. Each volunteer's weight and metabolic rate were recorded. This is an example of _____.

(A) an observational study
(B) a double-blind experiment
(C) a completely randomized experiment
(D) an experiment
(E) a block experiment

163. A researcher randomly selected 50 overweight volunteers to participate in a study on weight and energy consumption. Each volunteer's weight and metabolic rate were recorded. Then the researcher randomly assigned the subjects to one of two groups. One group received a metabolism-boosting supplement, and the other group received a placebo. Neither the subjects nor the researcher knew who received the treatment or the placebo. This is an example of _____.

(A) an observational study
(B) a double-blind experiment
(C) a completely randomized experiment
(D) an experiment
(E) a block experiment

164. A researcher randomly selected 50 online teachers to participate in a study on the effects of multitasking. Before the study began, teachers completed a survey that was used to determine a baseline rage index. Then the researcher randomly assigned teachers to one of two groups. One group was allowed to perform tasks such as grading assignments, responding to email, and teaching classes in one-hour intervals, while the other group had to perform tasks whenever they were received or demanded, with priority given to students requesting tutoring. After a period of one month, teachers again completed the survey to measure their rage index. Which of the statements describes this study?

   I. This is an experiment.
  II. This is an observational study.
 III. This is a matched-pairs design.

(A) I only
(B) I and III only
(C) II only
(D) I and II only
(E) III only

165. A researcher wishes to study the effects of diet on weight loss. He recruits 60 overweight subjects to participate in the study and records their gender. He also randomly assigns three different diet and exercise regimens to the subjects. Which of the following describes appropriate blocking techniques?

   I. Block on gender
  II. Block on diet
 III. Block on exercise

(A) I only
(B) I and III only
(C) II only
(D) I and II only
(E) III only

166. A study on the yield of broccoli wishes to compare application of two fertilizers: one provides phosphate (P) and the other nitrogen (N). The treatment levels are 0, 20, 40, and 60 pounds per acre.

(a) Describe how a completely randomized design would be implemented assuming that there are 64 plots of similar size with similar soil conditions.

(b) Describe how you would implement blocking if you had four fields in different locations and were concerned that soil conditions varied considerably.

(c) In the analysis of the results, what would the researcher wish to determine?

167. A study of multitasking causing increased rage is to be conducted with 40 female online teachers who have varying levels of experience.

(a) Describe how to assign experimental units to the following four treatment levels of multitasking: 0, 10, 20, and 30 task shifts per hour between answering email, grading assignments, tutoring students, answering and making phone contacts, responding to chats, and maintaining course web pages.

(b) Describe how you would implement blocking if there were 20 male and 20 female participants.

(c) Before each subject participates in the study, researchers use a survey to establish a baseline rage index. After participating in the multitasking treatment, subjects complete a survey for a new measurement of their rage index. What is this experimental design technique?

(d) If your results were inconclusive, what lurking variable could be having an effect on the outcomes? What could be done to control for this variation?

168. A study is conducted on female subjects to determine the effect of three types of therapy for pain: music, meditation, and exercise. The 48 females pre-screened for the study recorded pain levels of 8 or more on a scale of 0 to 10.

(a) If a leading pharmaceutical company wishes to show that its new pain medication is superior to these therapies, how would you design an experiment using the 48 female subjects?

(b) What if you suspected that there was a placebo effect associated with the new medication? What could you do?

(c) How could you control for researcher bias during the experiment?

169. A community college suspects that there is a relationship between student attendance and GPA. They select a cohort of six Introduction to Probability and Statistics classes with 50 students enrolled in each class.

(a) What observations should be made in order to determine a relationship between attendance and GPA?

(b) What type of study is this? Can you determine cause and effect?

(c) Suppose the college believes that teachers can increase the attendance levels in Introduction to Probability and Statistics by doing group work that would lead to higher GPAs. How could this be tested?

(d) If the outcome is that there is no significant difference in GPA between classes receiving group work and classes receiving traditional instruction, what could explain this outcome?

170. Psychologists observe and document a new disorder: Attachment Deficit (AD). They wish to determine whether there is a relationship between parenting styles and the incidence of AD. They extensively interview the parents of the children displaying the disorder to construct a set of characteristics.

(a) What type of study is this? Can they determine cause and effect?

(b) A pharmaceutical company suspects that one of its existing drugs to control psychotic episodes could be effective for AD if administered at a low dosage. Twenty parents agree to treat their AD children with the drug. What could be done if you suspect that there might be a placebo effect?

(c) Describe how to make the study double-blind. What are the reasons for doing double-blind experiments?

CHAPTER 5

Probability and Random Variables

171. Which of the following **best** describes a chance experiment?

(A) An activity where we can observe but can't predict outcomes
(B) Playing a game of poker
(C) Measuring the miles per gallon of a vehicle
(D) Picking a number between 1 and 10
(E) Picking a multi-choice answer without a clue

172. Individual outcomes of a random experiment are called _____.

(A) simple events
(B) random events
(C) experimental results
(D) experimental findings
(E) sample space

173. A sample space is _____.

(A) the set of experimental events from a number of trials
(B) the set of all events from a random experiment
(C) the set of all the experimental findings
(D) the set of all the experimental results
(E) the set of all possible outcomes of a random experiment

174. Which of the following statements **best** describes probability?

I. $0 \le P(E) \le 1$
II. The sum of the probabilities of all possible outcomes is 1
III. The probability of an event is the ratio of successes to failures

(A) I only
(B) I and III only
(C) II only
(D) I and II only
(E) III only

175. In a litter of five puppies, what is the probability of getting two males and three females, assuming an even chance for each gender?

(A) $\dfrac{1}{32}$

(B) 0.031

(C) $\dfrac{1}{16}$

(D) $\dfrac{5}{16}$

(E) $\dfrac{1}{4}$

176. In a litter of five puppies, what is the probability of getting two males and three females or three males and two females, assuming an even chance for each gender?

(A) 0.063

(B) $\dfrac{1}{16}$

(C) $\dfrac{5}{16}$

(D) $\dfrac{5}{8}$

(E) $\dfrac{1}{2}$

177. In a litter of five puppies, what is the probability of **not** getting two males and three females or three males and two females, assuming an even chance for each gender?

(A) $\dfrac{3}{8}$

(B) $\dfrac{5}{8}$

(C) $\dfrac{1}{2}$

(D) $\dfrac{11}{16}$

(E) 0.038

178. What is the probability of drawing a queen from a deck of cards, given that you have already drawn two kings and two queens and have not replaced them?

(A) $\dfrac{1}{52}$

(B) $\dfrac{1}{26}$

(C) $\dfrac{1}{48}$

(D) $\dfrac{1}{24}$

(E) 0.038

179. What is the probability of drawing a king or a queen, given that you have already drawn two kings and two queens and have not replaced them?

(A) $\dfrac{1}{12}$

(B) $\dfrac{1}{26}$

(C) $\dfrac{1}{13}$

(D) $\dfrac{2}{25}$

(E) 0.080

180. The probability of Bill serving an ace in tennis is 0.15, and the probability that he double faults is 0.25. What is the probability that Bill doesn't serve an ace or a double fault?

(A) 0.5

(B) 0.15

(C) 0.4

(D) 0.9

(E) 0.6

181. Which of the following **best** describes random variables?

   I. $0 \leq P(E) \leq 1$

   II. A value, discrete or continuous, that is assigned as an outcome of a random experiment

   III. A value, never continuous, that is an outcome of a random experiment

   (A) I only
   (B) I and III only
   (C) II only
   (D) I and II only
   (E) III only

182. Which of the following is **not** a random variable?

   (A) The roll of a fair die
   (B) The temperature in Vancouver, BC
   (C) The number of votes for different candidates in an election
   (D) The number of minutes in an hour
   (E) The heart rates of patients in Long Beach Memorial Hospital

183. What value completes the probability distribution below?

x	0	1	2	3	4	5
p(x)	1/5		1/5	1/6	1/6	1/8

   (A) $\dfrac{2}{15}$

   (B) $\dfrac{13}{105}$

   (C) $\dfrac{1}{2}$

   (D) $\dfrac{1}{5}$

   (E) $\dfrac{17}{120}$

184. Suppose you have an unfair die, one weighted so that the even numbers have twice the probability of showing as the odd numbers. Complete the probability distribution. Which statements are true?

x	1	2	3	4	5	6
p(x)						

   I. The distribution cannot be determined.

   II. $\dfrac{1}{9}, \dfrac{2}{9}, \dfrac{1}{9}, \dfrac{2}{9}, \dfrac{1}{9}, \dfrac{2}{9}$

   III. This is a continuous distribution.

(A) I only
(B) I and III only
(C) I and II only
(D) II only
(E) III only

185. Calculate the expected value and the variance of the following probability distribution.

x	1	2	3	4	5	6
p(x)	0.400	0.300	0.200	0.050	0.025	0.025

(A) $E(X) = 2.0$, $V(X) = 1.4$
(B) $E(X) = 2.1$, $V(X) = 1.0$
(C) $E(X) = 2.1$, $V(X) = 1.2$
(D) $E(X) = 2.1$, $V(X) = 1.4$
(E) $E(X) = 2.0$, $V(X) = 1.0$

186. The probability distribution for the number of incidents occurring in a year is shown below. Calculate the expected value and the variance.

Incidents, x	0	1	2	3	4	5
p(x)	0.1353	0.2707	0.2707	0.1804	0.0902	0.0527

(A) $E(X) = 2.0$, $V(X) = 2.0$
(B) $E(X) = 1.9$, $V(X) = 1.8$
(C) $E(X) = 2.0$, $V(X) = 1.8$
(D) $E(X) = 2.0$, $V(X) = 1.4$
(E) $E(X) = 1.9$, $V(X) = 1.4$

187. The probability distribution for the frequency of an event occurring in five trials is shown below. Calculate the expected value and the standard deviation.

Frequency, x	0	1	2	3	4	5
p(x)	0.16807	0.36015	0.30870	0.13230	0.02835	0.00243

(A) $E(X) = 1.25$, $SD = 1.0$
(B) $E(X) = 1.5$, $SD = 1.05$
(C) $E(X) = 1.5$, $SD = 1.1$
(D) $E(X) = 1.5$, $SD = 1.0$
(E) $E(X) = 1.25$, $SD = 1.05$

188. The probability distribution for the number of defects found in a sample is shown below. Calculate the expected value and the standard deviation.

Defects, x	0	1	2	3
p(x)	0.3385	0.4481	0.1883	0.0251

(A) $E(X) = 0.9$, $SD = 0.8$
(B) $E(X) = 0.9$, $SD = 0.6$
(C) $E(X) = 0.9$, $SD = 0.7$
(D) $E(X) = 1.0$, $SD = 0.6$
(E) $E(X) = 1.0$, $SD = 0.7$

189. Which of the following probability distributions matches the histogram?

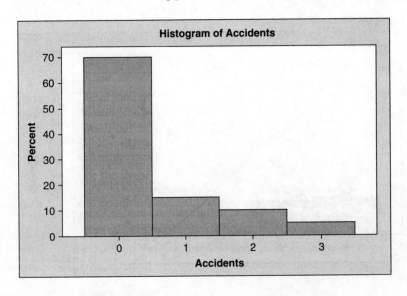

(A)

Accidents, x	0	1	2	3
p(x)	0.7	0.2	0.1	0.5

(B)

Accidents, x	0	1	2	3
p(x)	0.7	0.2	0.1	0.5

(C)

Accidents, x	0	1	2	3
p(x)	7.0	1.5	1.0	0.5

(D)

Accidents, x	0	1	2	3
p(x)	70	15	10	5

(E)

Accidents, x	0	1	2	3
p(x)	0.70	0.15	0.10	0.05

190. What should be the height of the missing bar in the histogram?

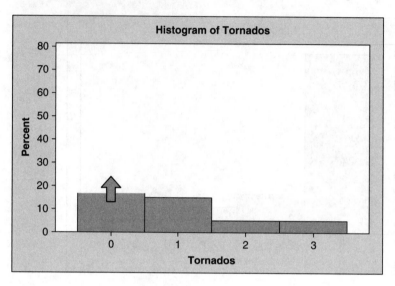

(A) 75
(B) 0.7
(C) 70
(D) 60
(E) 50

191. Suppose a cell phone company manufactures cell phones and tablets in the following four states: Arizona, California, Washington, and Minnesota. The table below shows the percentage of total output by state, and within each state, the percentage output of each product. What is the probability that a randomly selected product will be a tablet manufactured in Minnesota? (Assume that all products are shipped from one distribution center.)

Arizona		California		Washington		Minnesota	
25%		45%		25%		5%	
Phone	Tablet	Phone	Tablet	Phone	Tablet	Phone	Tablet
78%	22%	56%	44%	86%	14%	92%	8%

(A) 0.05
(B) 0.08
(C) 0.004
(D) 0.5
(E) 0.04

192. Suppose a cell phone company manufactures cell phones and tablets in the following four states: Arizona, California, Washington, and Minnesota. The table below shows the percentage of total output by state, and within each state, the percentage output of each product. What is the probability of selecting at random a cell phone made in California while you're shopping for a phone? (Assume that all products are shipped from one distribution center.)

Arizona		California		Washington		Minnesota	
25%		45%		25%		5%	
Phone	Tablet	Phone	Tablet	Phone	Tablet	Phone	Tablet
78%	22%	56%	44%	86%	14%	92%	8%

(A) 0.181
(B) 0.129
(C) 0.144
(D) 0.252
(E) 0.356

193. Suppose a cell phone company manufactures cell phones and tablets in the following four states: Arizona, California, Washington, and Minnesota. The table below shows the percentage of total output by state, and within each state, the percentage output of each product. What is the probability of selecting at random a cell phone made in California or Arizona while you're shopping for a phone? (Assume that all products are shipped from one distribution center.)

Arizona		California		Washington		Minnesota	
25%		45%		25%		5%	
Phone	Tablet	Phone	Tablet	Phone	Tablet	Phone	Tablet
78%	22%	56%	44%	86%	14%	92%	8%

(A) 0.191
(B) 0.130
(C) 0.631
(D) 0.447
(E) 0.808

194. What is the probability that a single card drawn from a standard deck of cards (jokers removed) will be a jack or a heart?

(A) $\dfrac{15}{52}$

(B) $\dfrac{1}{52}$

(C) $\dfrac{4}{221}$

(D) $\dfrac{17}{52}$

(E) $\dfrac{4}{13}$

195. What is the probability of drawing a jack first and a three second from a standard deck of cards (jokers removed), when two cards are drawn without replacement?

(A) $\dfrac{1}{169}$

(B) $\dfrac{2}{13}$

(C) $\dfrac{16}{663}$

(D) $\dfrac{2}{663}$

(E) $\dfrac{1}{338}$

196. What is the probability that you and your friend choose the same number between 1 and 20?

(A) 0.5

(B) 0.05

(C) 0.0025

(D) 0.053

(E) 0.005

197. What is the probability of rolling at least one six on a pair of fair dice?

(A) $\dfrac{11}{36}$

(B) $\dfrac{1}{3}$

(C) $\dfrac{1}{6}$

(D) $\dfrac{1}{15}$

(E) $\dfrac{1}{16}$

198. Consider the following table. What is the probability of a sum of five when a fair four-sided die and a fair five-sided die are rolled?

Sums	1	2	3	4
1	2	3	4	5
2	3	4	5	6
3	4	5	6	7
4	5	6	7	8
5	6	7	8	9

(A) $\dfrac{4}{25}$

(B) $\dfrac{1}{4}$

(C) $\dfrac{1}{5}$

(D) $\dfrac{4}{5}$

(E) $\dfrac{4}{9}$

199. Consider the following table. What is the probability of a sum of two or nine when a fair four-sided die and a fair five-sided die are rolled?

Sums	1	2	3	4
1	2	3	4	5
2	3	4	5	6
3	4	5	6	7
4	5	6	7	8
5	6	7	8	9

(A) $\dfrac{1}{10}$

(B) $\dfrac{2}{25}$

(C) $\dfrac{1}{8}$

(D) $\dfrac{2}{9}$

(E) $\dfrac{2}{5}$

200. Consider the following table. What is the probability of rolling at least one three when a fair four-sided die and a fair five-sided die are rolled?

Sums	1	2	3	4
1	2	3	4	5
2	3	4	5	6
3	4	5	6	7
4	5	6	7	8
5	6	7	8	9

(A) $\dfrac{9}{20}$

(B) $\dfrac{2}{5}$

(C) $\dfrac{8}{25}$

(D) $\dfrac{1}{2}$

(E) $\dfrac{4}{9}$

201. Which of these statements is true about continuous random variables and their probability distributions?
    I. The distribution is modeled by a probability distribution function that is non-negative.
    II. The probability of an individual event E is calculated by $P(X = E)$.
    III. The total area under the curve is 1.

    (A) I only
    (B) I and II only
    (C) I and III only
    (D) II only
    (E) III only

202. Which of the statements is true about the normal probability distribution?
    I. $\mu = 0$ and $\sigma = 1$ only.
    II. It is symmetric about μ with tails extending to positive and negative infinity.
    III. The total area under the curve cannot be determined because the tails are asymptotic to the horizontal axis.

    (A) I only
    (B) I and II only
    (C) I and III only
    (D) II only
    (E) III only

203. Which of the statements is true about the necessary conditions for using the normal probability distribution?
    I. If $\mu = 0$ and $\sigma = 1$, you know that the distribution is normal.
    II. The normal distribution can be used at any time regardless of a population's distribution.
    III. When it is known that a distribution is approximately normal, the normal distribution may be used.

    (A) I only
    (B) I and II only
    (C) I and III only
    (D) II only
    (E) III only

204. In the standard normal distribution, what is the probability that $z \geq 1.25$?

    (A) 0.8925
    (B) 0.8962
    (C) 0.8944
    (D) 0.1056
    (E) 0.1038

205. In the standard normal distribution, what is the probability that $-2.95 < z \leq 0.95$?

(A) 0.0016
(B) 0.8289
(C) 0.8305
(D) 0.8273
(E) 0.8276

206. The average height of women between the ages 30–39 is 163.2 centimeters with a standard deviation of 9.3 centimeters. Find the probability that a woman in this age group is over 160 centimeters if it is known that the distribution is approximately normal.

(A) 0.3669
(B) 0.3655
(C) 0.6331
(D) 0.6360
(E) 0.3660

207. The average weight of men between the ages 40–49 is 202.3 pounds with a standard deviation of 50.7 pounds. Find the probability that a man in this age group is under 180 pounds if it is known that the distribution is approximately normal.

(A) 0.6700
(B) 0.6331
(C) 0.3301
(D) 0.3300
(E) 0.3669

208. The average weight of men between the ages 20–29 is 188.3 pounds with a standard deviation of 66.4 pounds. The average weight of women between the ages 20–29 is 155.9 pounds with a standard deviation of 60.3 pounds. If it is known that the distributions are approximately normal, which of the statements is true?

   I. A woman weighing 215 pounds is more likely than a man weighing 255 pounds.
  II. A man weighing 255 pounds is more likely than a woman weighing 215 pounds.
 III. They are equally likely.

(A) I only
(B) I and II only
(C) I and III only
(D) II only
(E) III only

209. If SAT scores are normally distributed with a mean of 500 and a standard deviation of 100, what minimum score is needed to ensure that you are in the top 7%?

(A) 640
(B) 500
(C) 645
(D) 650
(E) 648

210. The average body mass index value of children 9 years old is 18.4 with a standard deviation of 4.5. If it is known that the distributions are approximately normal, what is the index for children at the 82nd percentile?

(A) 22.6
(B) 22.0
(C) 23.0
(D) 0.92
(E) 0.82

211. Which of these statements is true about the law of large numbers?
   I. One hundred trials will be enough to determine the true proportion in the population.
  II. Over time the proportion of successes in a simulation approaches the true proportion in the population.
 III. The relative frequency of successes from just a few trials accurately approximates the true probability of success.

(A) I only
(B) I and II only
(C) II only
(D) I and III only
(E) III only

212. What would be an appropriate way to assign digits in order to simulate flipping an unfair coin with $P(H) = 0.3$?

(A) Assign digits 1, 2, 3 to be a head. Assign digits 4–9 to be a tail.
(B) Assign digits 0, 1, 2, 3 to be a head. Assign digits 4–9 to be a tail.
(C) Assign digits 0, 1, 2, 3 to be a tail. Assign digits 4–9 to be a head.
(D) Assign digits 0, 1, 2 to be a tail. Assign digits 3–9 to be a head.
(E) Assign digits 0, 1, 2 to be a head. Assign digits 3–9 to be a tail.

213. Students in a biomedical statistics course take two tests before a final. The statistics from the first test for three different classes (May, June, and July) are shown. What would be the average and the standard deviation of the sum of the test scores?

	Test Statistics	
Month	Average	Standard Deviation
May	17	1.25
June	16	1.85
July	18	2.35

(A) Mean of the sum: 51. Standard deviation: 10.51.
(B) Mean of the sum: 51. Standard deviation: 5.45.
(C) Mean of the sum: 51. Standard deviation: 3.24.
(D) Mean of the sum: 51. Standard deviation: 2.33.
(E) Mean of the sum: 50. Standard deviation: 3.24.

214. Which of the statements is true about adding two random variables X and Y?
    I. The mean of the new distribution for $X + Y$ is $\mu_X \mu_Y$ and the variance is $\sigma_X^2 + \sigma_Y^2$.
    II. The mean of the new distribution for $X + Y$ is $\mu_X + \mu_Y$ and the variance is $\sigma_X^2 \sigma_Y^2$.
    III. The mean of the new distribution for $X + Y$ is $\mu_X + \mu_Y$.

(A) I only
(B) I and III only
(C) I and II only
(D) II only
(E) III only

215. Which of the statements is true about transforming random variables?
    I. A random variable X added to a constant a has a mean of $a + \mu_X$ and a variance of σ_X^2.
    II. A random variable X multiplied by a constant b has a mean of $b\mu_X$ and a variance of $b^2 \sigma_X^2$.
    III. A random variable X multiplied by a constant b has a mean of $b\mu_X$ and a variance of $b\sigma_X^2$.

(A) I only
(B) I and III only
(C) II only
(D) I and II only
(E) III only

216. Which of the following represents the probability that someone who works full time has more than $5,000 in credit card debt?

(A) P(full time and credit card debt over $5,000)
(B) P(full time or credit card debt over $5,000)
(C) P(full time | credit card debt over $5,000)
(D) P(credit card debt over $5,000 | full time)
(E) P(full time)*P(credit card debt over $5,000)

217. Given the following probability distribution, find $P(1 \leq X \leq 4)$.

x	1	2	3	4	5
p(x)	0.12	0.01	0.07	0.50	0.30

(A) 0.01
(B) 0.08
(C) 0.20
(D) 0.58
(E) 0.70

218. In a statistics course, the probability that a randomly selected student has taken a calculus course is 0.14, while the probability that a randomly selected student drives to campus is 0.48. If the two events are independent, what is the probability that a randomly selected student in this statistics course has taken a calculus course and drives to campus?

(A) 0.07
(B) 0.14
(C) 0.34
(D) 0.48
(E) 0.62

219. The dataset below is based on a survey of 100 randomly selected travelers who were asked, "How many trips per year do you typically take?"

	2 or Fewer	3 to 5 Trips	6 to 8 Trips	9 or More Trips
Male	17	10	20	3
Female	19	21	8	2

What is the probability that a randomly selected survey participant is male or takes two or fewer trips per year?

(A) 0.17
(B) 0.18
(C) 0.36
(D) 0.69
(E) 0.86

220. In a contest at a state fair, participants tried to see how many times in a row they could throw a football through a determined space until they missed. The company running the contest compiled the following data during the entire fair.

Number of Throws	1	2	3	4	5	6	7	8	9
Number of Contestants	155	220	380	187	121	85	60	15	4

(a) Construct a probability distribution based on these data.
(b) If a participant was randomly selected, how many throws would you expect him or her to make before missing? Justify your answer.

221. A Sonoma County blood bank ran a blood drive on the northern California coast and compiled the following frequency tabulation from the donors.

Rh-factor	Blood Type			
	0	A	B	AB
Positive	110	98	26	9
Negative	20	18	6	3

(a) Calculate the probabilities for each blood type based on these donors.
(b) Calculated the following probabilities based on these donors:
P (type A)
P (Rh-pos)
P (not type O)
P (type O or type AB)
P (type O or Rh-neg)
(c) Using this sample, how many donors of type O-negative, the universal blood type, would they expect in a Marin County blood drive that drew 467 donors?

222. Mars, Inc. manufactures plain M&M's in six colors: yellow, green, red, brown, blue, and orange. Your grandmother keeps a large glass bowl of plain M&M's on the dining table. M&M's are, in theory, manufactured and packaged with a uniform distribution. You love to close your eyes, reach in, and select one M&M at a time. Because of your resolution to lose weight, you have also resolved not to eat them. To squelch your appetite for M&M's, you simulate the process of drawing of individual M&M's from the bowl, one at a time, using a random number table.

(a) Assign the digits from a random number table appropriately for a simulation of drawing one M&M at a time from your grandmother's bowl based on the assumption of a uniform distribution of the six colors.
(b) Using the random number table provided here and starting at the indicated digit, tabulate 30 trials.

Part of a Table of Random Numbers			
61424	2⬇419	86546	00517
90222	➡7993	04952	66762
50349	71146	97668	86523
85676	10005	08216	25906
02429	19761	15370	43882
90519	61988	40164	15815
20631	88967	19660	89624
89990	78733	16447	27932

(c) Calculate the empirical (statistical) probabilities based on the simulation. What are the classical (theoretical) probabilities based on a uniform distribution?
(d) What does the law of large numbers state about empirical probabilities in a simulation?

223. The greater spot-nosed monkey is the smallest Old World monkey found on the west coast of Africa. Its head and body length average 385.0 mm with a standard deviation of 21.7 mm. Its average mass is 1005.0 g with a standard deviation of 81.6 g. Assume that its lengths and masses are normally distributed.

(a) Two greater spot-nosed monkeys brought to my mother had masses of 805.0 g and 1195.0 g. Find the probability that a greater spot-nosed monkey selected at random has a mass between 805.0 and 1195.0 g.

(b) The two monkeys' head and body lengths measured 405.0 mm and 375.0 mm. Find the probability that a greater spot-nosed monkey is longer than 405.0 or shorter than 375.0.

(c) Which is more likely, randomly selecting a greater spot-nosed monkey that is 405.0 mm ±0.5 mm long or one with a mass of 1195.0 g ±0.5 g?

224. In 2009, SAT scores for mathematics averaged 515 with a standard deviation of 116. Assume the scores are normally distributed.

(a) A score of 500 would be at what percentile? At what percentile would a score of 720 be?

(b) If in 1999, SAT scores for mathematics were on average of 511 and a standard deviation of 136, a score of 500 would be at what percentile? At what percentile would a score of 720 be?

(c) Which year had higher scores above the 90th percentile? Which year had higher scores below the 25th percentile?

(d) If a news article reported that scores in 1999 were better than scores in 2009, would you agree? Why or why not?

225. Two biomedical statistics classes offered in different semesters produced the following scores on the first quiz.

Courses	
#1	**#2**
14/20	16/20
16/20	16/20
19/20	13/20
15/20	13/20
18/20	16/20
17/20	15/20
15/20	16/20
19/20	16/20
15/20	16/20
20/20	17/20
18/20	17/20
18/20	
16/20	
20/20	
18/20	
15/20	
19/20	
17/20	

(a) Let the scores from Course #1 be the random variable X and the scores from Course #2 be the random variable Y. Calculate the mean and the standard deviation from each class.

(b) Assuming that the distribution for this quiz is normal, find the score of the 90th percentile in Course #1 and Course #2. If this is an indication of the level of difficulty of these courses, in which class is it easier to get a 90% or above?

(c) What would be the mean and standard deviation of the total score $X + Y$?

CHAPTER 6

Binomial Distribution, Geometric Distribution, and Sampling

226. A binomial random variable X counts the number of successes in n trials. What are the mean and standard deviation of X if the probability of failure is 0.29 and is repeated 145 times?

(A) 42.05; 5.46

(B) 102.95; 5.46

(C) 42.05; 29.86

(D) 102.95; 29.86

(E) It is not possible to find the mean or the standard deviation of a binomial experiment using the probability of failure

227. A binomial experiment has $n = 75$ trials. The probability of success for each trial is 0.3. Let X be the count of successes of the event during the 75 trials. What are μ_x and σ_X?

(A) 22.5; 15.75

(B) 22.5; 3.97

(C) 52.5; 15.75

(D) 52.5; 3.97

(E) The standard deviation cannot be determined from the given information

228. What are the mean and the standard deviation of a binomial experiment that occurs with probability of success of 0.3 and is repeated 150 times?

(A) 45; 5.6

(B) 45; 31.5

(C) 105; 5.6

(D) 105; 31.5

(E) The standard deviation cannot be determined from the given information

229. A binomial experiment has $n = 50$ trials. The probability of failure for each trial is 0.4. Let X be the count of successes of the event during the 50 trials. What are μ_x and σ_X?

(A) 30; 12

(B) 20; 12

(C) 30; 3.5

(D) 20; 3.5

(E) It is not possible to find the mean or the standard deviation of a binomial experiment using the probability of failure

230. A random variable X counts the number of successes in 15 independent trials. There is the same probability of success for each trial. If the probability of exactly 1 success in the 15 trials is $\binom{15}{1}(0.3)^1(0.7)^{14}$, what is μ_x?

(A) 4.5

(B) 10.5

(C) 3.15

(D) 1.77

(E) 1.35

231. A random variable X counts the number of successes in 20 independent trials. There is the same probability of success for each trial. If the probability of exactly 1 success in the 20 trials is $\binom{20}{1}(0.4)^1(0.6)^{19}$, what is σ_X?

(A) 8

(B) 12

(C) 4.8

(D) 2.2

(E) 3.2

232. A binomial random variable X counts the number of successes in 50 trials. If $\mu_x = 20$, what is the probability of success for a single trial?

(A) 0.4

(B) 0.84

(C) 0.6

(D) 0.16

(E) The probability of success cannot be determined from the given information

233. A binomial random variable X counts the number of successes in 40 trials. If $\mu_x = 30$, what is the probability of failure for a single trial?

(A) 0.75
(B) 0.25
(C) 0.56
(D) 0.44
(E) The probability of failure cannot be determined from the given information

234. A random variable X counts the number of successes in 25 independent trials. There is the same probability of success for each trial. If the probability of exactly 1 success in the 50 trials is $\binom{50}{1}(0.2)^1(0.8)^{49}$, what is the probability of success on the first trial?

(A) 0.2
(B) 0.8
(C) $\binom{50}{1}(0.2)^1(0.8)^{49}$

(D) $\binom{50}{0}(0.2)^0(0.8)^{50}$

(E) $\binom{50}{1}(0.2)^{49}(0.8)^1$

235. A random variable X counts the number of successes in 30 independent trials. There is the same probability of success for each trial. If the probability of exactly 4 successes in the 30 trials is $\binom{30}{4}(0.2)^4(0.8)^{26}$, what is the probability of failure on the fourth trial?

(A) 0.2
(B) 0.8
(C) $\binom{30}{4}(0.2)^4(0.8)^{26}$

(D) 0.32
(E) $1-\left[\binom{30}{1}(0.2)^1(0.8)^{29}\right]$

236. A binomial experiment consists of 40 trials. If the random variable X counts the number of successes and the probability of success for any one trial is 0.3, write the mathematical expression you would need to evaluate to find $P(X = 17)$. Do not evaluate.

(A) $\binom{40}{17}(0.3)^{17}(0.7)^{23}$

(B) $\binom{40}{23}(0.3)^{40}(0.7)^{23}$

(C) $\binom{40}{23}(0.3)^{23}(0.7)^{17}$

(D) $\binom{40}{17}(0.3)^{23}(0.7)^{17}$

(E) $\binom{40}{17}(0.3)^{40}(0.7)^{17}$

237. A fair coin is flipped 100 times. Which of the following would give the exact probability of getting heads at least 80 times?

(A) $1 - \left[\binom{100}{0}\left(\frac{1}{2}\right)^{0}\left(\frac{1}{2}\right)^{100} + \binom{100}{1}\left(\frac{1}{2}\right)^{1}\left(\frac{1}{2}\right)^{99} + \cdots + \binom{100}{79}\left(\frac{1}{2}\right)^{79}\left(\frac{1}{2}\right)^{21}\right]$

(B) $1 - \left[\binom{100}{0}\left(\frac{1}{2}\right)^{0}\left(\frac{1}{2}\right)^{100} + \binom{100}{1}\left(\frac{1}{2}\right)^{1}\left(\frac{1}{2}\right)^{99} + \cdots + \binom{100}{80}\left(\frac{1}{2}\right)^{80}\left(\frac{1}{2}\right)^{20}\right]$

(C) $\binom{100}{81}\left(\frac{1}{2}\right)^{81}\left(\frac{1}{2}\right)^{19} + \binom{100}{82}\left(\frac{1}{2}\right)^{82}\left(\frac{1}{2}\right)^{18} + \cdots + \binom{100}{100}\left(\frac{1}{2}\right)^{100}\left(\frac{1}{2}\right)^{0}$

(D) $\binom{100}{0}\left(\frac{1}{2}\right)^{0}\left(\frac{1}{2}\right)^{100} + \binom{100}{1}\left(\frac{1}{2}\right)^{1}\left(\frac{1}{2}\right)^{99} + \cdots + \binom{100}{80}\left(\frac{1}{2}\right)^{80}\left(\frac{1}{2}\right)^{20}$

(E) $\binom{100}{0}\left(\frac{1}{2}\right)^{0}\left(\frac{1}{2}\right)^{100} + \binom{100}{1}\left(\frac{1}{2}\right)^{1}\left(\frac{1}{2}\right)^{99} + \cdots + \binom{100}{79}\left(\frac{1}{2}\right)^{79}\left(\frac{1}{2}\right)^{21}$

238. A fair coin is flipped 100 times. Which of the following would give the exact probability of getting heads at most 80 times?

(A) $1 - \left[\binom{100}{0}\left(\frac{1}{2}\right)^0\left(\frac{1}{2}\right)^{100} + \binom{100}{1}\left(\frac{1}{2}\right)^1\left(\frac{1}{2}\right)^{99} + \cdots + \binom{100}{79}\left(\frac{1}{2}\right)^{79}\left(\frac{1}{2}\right)^{21} \right]$

(B) $1 - \left[\binom{100}{0}\left(\frac{1}{2}\right)^0\left(\frac{1}{2}\right)^{100} + \binom{100}{1}\left(\frac{1}{2}\right)^1\left(\frac{1}{2}\right)^{99} + \cdots + \binom{100}{80}\left(\frac{1}{2}\right)^{80}\left(\frac{1}{2}\right)^{20} \right]$

(C) $\binom{100}{81}\left(\frac{1}{2}\right)^{81}\left(\frac{1}{2}\right)^{19} + \binom{100}{82}\left(\frac{1}{2}\right)^{82}\left(\frac{1}{2}\right)^{18} + \cdots + \binom{100}{100}\left(\frac{1}{2}\right)^{100}\left(\frac{1}{2}\right)^{0}$

(D) $\binom{100}{0}\left(\frac{1}{2}\right)^0\left(\frac{1}{2}\right)^{100} + \binom{100}{1}\left(\frac{1}{2}\right)^1\left(\frac{1}{2}\right)^{99} + \cdots + \binom{100}{80}\left(\frac{1}{2}\right)^{80}\left(\frac{1}{2}\right)^{20}$

(E) $\binom{100}{0}\left(\frac{1}{2}\right)^0\left(\frac{1}{2}\right)^{100} + \binom{100}{1}\left(\frac{1}{2}\right)^1\left(\frac{1}{2}\right)^{99} + \cdots + \binom{100}{79}\left(\frac{1}{2}\right)^{79}\left(\frac{1}{2}\right)^{21}$

239. A fair die is to be rolled 12 times. What is the probability of getting at most two fives?

(A) $\binom{12}{0}\left(\frac{1}{6}\right)^0\left(\frac{5}{6}\right)^{12} + \binom{12}{1}\left(\frac{1}{6}\right)^1\left(\frac{5}{6}\right)^{11} + \binom{12}{2}\left(\frac{1}{6}\right)^2\left(\frac{5}{6}\right)^{10}$

(B) $\binom{12}{0}\left(\frac{1}{6}\right)^0\left(\frac{5}{6}\right)^{12} + \binom{12}{1}\left(\frac{1}{6}\right)^1\left(\frac{5}{6}\right)^{11}$

(C) $1 - \left[\binom{12}{0}\left(\frac{1}{6}\right)^0\left(\frac{5}{6}\right)^{12} + \binom{12}{1}\left(\frac{1}{6}\right)^1\left(\frac{5}{6}\right)^{11} + \binom{12}{2}\left(\frac{1}{6}\right)^2\left(\frac{5}{6}\right)^{10} \right]$

(D) $1 - \left[\binom{12}{0}\left(\frac{1}{6}\right)^0\left(\frac{5}{6}\right)^{12} + \binom{12}{1}\left(\frac{1}{6}\right)^1\left(\frac{5}{6}\right)^{11} \right]$

(E) $\binom{12}{0}\left(\frac{1}{6}\right)^0\left(\frac{5}{6}\right)^{12} + \binom{12}{1}\left(\frac{1}{6}\right)^1\left(\frac{5}{6}\right)^{11} + \cdots + \binom{12}{5}\left(\frac{1}{6}\right)^5\left(\frac{5}{6}\right)^{7}$

240. A fair die is to be rolled 10 times. What is the probability of getting more than one three?

(A) $\binom{10}{0}\left(\frac{1}{6}\right)^0\left(\frac{5}{6}\right)^{10}+\binom{10}{1}\left(\frac{1}{6}\right)^1\left(\frac{5}{6}\right)^9$

(B) $1-\left[\binom{10}{0}\left(\frac{1}{6}\right)^0\left(\frac{5}{6}\right)^{10}+\binom{10}{1}\left(\frac{1}{6}\right)^1\left(\frac{5}{6}\right)^9\right]$

(C) $1-\left[\binom{10}{0}\left(\frac{1}{6}\right)^0\left(\frac{5}{6}\right)^{10}\right]$

(D) $\binom{10}{0}\left(\frac{1}{6}\right)^0\left(\frac{5}{6}\right)^{10}$

(E) $\binom{10}{0}\left(\frac{1}{6}\right)^0\left(\frac{5}{6}\right)^{10}+\binom{10}{1}\left(\frac{1}{6}\right)^1\left(\frac{5}{6}\right)^9+\cdots+\binom{10}{3}\left(\frac{1}{6}\right)^3\left(\frac{5}{6}\right)^7$

241. Historically, 65% of employees at a company receive a positive review after their first year of employment. If 300 employees who have been employed for more than a year are randomly selected, what is the approximate probability fewer than 200 received a positive review after their first year of employment?

(A) $\binom{300}{200}(0.65)^{200}(0.35)^{100}$

(B) $\binom{300}{199}(0.65)^{199}(0.35)^{101}$

(C) $P\left(Z<\dfrac{200-195}{\sqrt{300(0.65)(0.35)}}\right)$

(D) $P\left(Z<\dfrac{200-195}{\sqrt{200(0.65)(0.35)}}\right)$

(E) $P\left(Z<\dfrac{195-200}{\sqrt{300(0.65)(0.35)}}\right)$

242. A six-sided die numbered 1–6 is weighted so that the probability of getting a 3 on a single roll is ¼. If the die is rolled 80 times, what is the approximate probability of getting a 3 on more than 50 of the rolls?

(A) $\dbinom{80}{50}\left(\dfrac{1}{4}\right)^{80}\left(\dfrac{3}{4}\right)^{30}$

(B) $1-\left[\dbinom{80}{50}\left(\dfrac{1}{4}\right)^{80}\left(\dfrac{3}{4}\right)^{30}\right]$

(C) $P\left(Z<\dfrac{50-20}{\sqrt{80\left(\dfrac{1}{4}\right)\left(\dfrac{3}{4}\right)}}\right)$

(D) $P\left(Z<\dfrac{50-20}{\sqrt{50\left(\dfrac{1}{4}\right)\left(\dfrac{3}{4}\right)}}\right)$

(E) $P\left(Z<\dfrac{20-50}{\sqrt{80\left(\dfrac{1}{4}\right)\left(\dfrac{3}{4}\right)}}\right)$

243. A binomial experiment has $n = 40$ trials. The probability of success on each trial is 0.3. What is the probability the first success will occur on the 15th trial?

(A) $\dbinom{40}{15}(0.3)^{15}(0.7)^{25}$

(B) $\dbinom{40}{15}(0.3)^{25}(0.7)^{15}$

(C) $(0.3)(0.7)^{15}$

(D) $(0.3)(0.7)^{14}$

(E) $(0.7)(0.3)^{15}$

244. A binomial experiment has $n = 25$ trials. The probability of failure on each trial is 0.6. What is the probability the first success will occur on the 15th trial?

(A) $\binom{25}{15}(0.6)^{15}(0.4)^{25}$

(B) $\binom{25}{15}(0.4)^{15}(0.6)^{25}$

(C) $(0.4)(0.6)^{15}$

(D) $(0.4)(0.6)^{14}$

(E) $(0.6)(0.4)^{15}$

245. It is known that 80% of the items in a population have a given property. If all possible samples of size 60 are taken from this population, which of the following would describe the sampling distribution of $\hat{p}$, the sample proportion of items with the given property?

(A) The sampling distribution is skewed right with a mean of 0.8 and a standard deviation of 0.31.

(B) The sampling distribution is skewed left with a mean of 0.1 and a standard deviation of 0.05.

(C) The sampling distribution is approximately normal with a mean of 0.8 and a standard deviation of 0.05.

(D) The sampling distribution is approximately normal with a mean of 0.1 and a standard deviation of 0.31.

(E) The conditions are not met to make any statements about the sampling distribution.

246. Sixty trials of a binomial random variable X are conducted where X counts the number of successes in the 60 trials. If the probability of failure for any one trial is 0.2, write the mathematical expression you would need to evaluate to find $P(X = 10)$.

(A) $\binom{60}{10}(0.2)^{60}(0.8)^{10}$

(B) $1 - \left[\binom{60}{10}(0.2)^{60}(0.8)^{50}\right]$

(C) $\binom{60}{10}(0.8)^{10}(0.2)^{50}$

(D) $\binom{60}{10}(0.8)^{60}(0.2)^{10}$

(E) $\binom{60}{10}(0.2)^{10}(0.8)^{50}$

247. Consider a binomial random variable X that counts the number of successes in 100 trials. If $\mu_X = 15$, find the probability of exactly 21 successes in the 100 trials.

(A) 0.973

(B) 0.027

(C) 0.098

(D) 0.902

(E) The probability cannot be determined from the given information

248. A student taking a 20-question multiple-choice quiz guesses each answer randomly from five choices. Each correct answer is worth one point, an incorrect answer is worth no points, and the quiz score is reported as a percentage. Which expression represents the probability the student will score a 60% on the quiz?

(A) $\left(\dfrac{1}{5}\right)^{12}\left(\dfrac{4}{5}\right)^{8}$

(B) $(0.6)\left(\dfrac{1}{5}\right)^{12}\left(\dfrac{4}{5}\right)^{8}$

(C) $\dbinom{20}{12}\left(\dfrac{1}{5}\right)^{12}\left(\dfrac{4}{5}\right)^{8}$

(D) $\dbinom{60}{20}\left(\dfrac{1}{5}\right)^{12}\left(\dfrac{4}{5}\right)^{8}$

(E) $\left(\dfrac{1}{5}\right)^{8}\left(\dfrac{4}{5}\right)^{12}$

249. A random variable X counts the number of successes in 16 independent trials. The probability any one trial is successful is 0.2. What is the probability of exactly five successful trials?

(A) 0.2
(B) 0.88
(C) 0.8
(D) 0.92
(E) 0.12

250. A random variable X counts the number of successes in 20 independent trials. The probability that any one trial is unsuccessful is 0.42. What is the probability of exactly eight successful trials?

(A) 0.17
(B) 0.05
(C) 0.52
(D) 0.08
(E) 0.42

251. The number of successes in ten trials is counted by the binomial random variable X. If the probability that the fifth trial is successful is 0.8, what is the probability the sixth trial will be a success?

(A) 0.09
(B) 0.12
(C) 0.64
(D) 0.2
(E) 0.8

252. The number of successes in 15 trials is counted by the binomial random variable X. If the probability that the third trial is a failure is 0.3, what is the probability the fourth trial will be successful?

(A) 0.7
(B) 0.3
(C) 0.22
(D) 0.13
(E) 0.09

253. A binomial experiment consists of 12 repeated trials. If the probability that the first success occurs on the second trial is 0.25, what is the probability of exactly 7 successful trials?

(A) 0.25
(B) 0.75
(C) 0.19
(D) 0.01
(E) The probability cannot be determined from the given information

254. The probability of two successes in two trials of a binomial experiment is 0.09. If the same experiment is repeated for 10 trials, what is the probability of exactly 3 successes in those 10 trials?

(A) $\binom{10}{3}(0.09)^3(0.91)^7$

(B) $\binom{10}{2}(0.09)^2(0.91)^8$

(C) $\binom{10}{3}(0.3)^3(0.7)^7$

(D) $\binom{10}{2}(0.3)^2(0.7)^8$

(E) $\binom{10}{3}(0.09)^7(0.91)^3$

255. A sample of size 35 is drawn from a population with a skewed right distribution, and the sample mean is calculated. If the mean of the population is 15 with a standard deviation of 2.1, what is the probability the sample mean will be larger than 14?

(A) $P\left(z > \dfrac{14-15}{2.1/\sqrt{35}}\right)$

(B) $P\left(z > \dfrac{14-15}{2.1}\right)$

(C) $P\left(z > \dfrac{15-14}{2.1/\sqrt{35}}\right)$

(D) $P\left(z > \dfrac{15-14}{2.1}\right)$

(E) The probability cannot be determined using typical methods since the shape of the sampling distribution is unknown

256. A large population has a distribution that is heavily skewed to the right with a mean of 10 and a standard deviation of 2.5. If all possible samples of size 45 are taken, which of the following statements will be true about the distribution of sample means?

(A) The distribution will be approximately normal with a mean of 10 and a standard deviation of 2.5.

(B) The distribution will be skewed to the right with a mean of 10 and a standard deviation of 2.5.

(C) The shape of the distribution cannot be determined, but the mean will be 10 with a standard deviation of 2.5.

(D) The distribution will be skewed to the right with a mean of 10 and a standard deviation of 0.37.

(E) The distribution will be approximately normal with a mean of 10 and a standard deviation of 0.37.

257. A sample of size 8 is drawn from a population with an unknown distribution, and the sample mean is calculated. If the mean of the population is 27 with a standard deviation of 8.5, what is the probability the sample mean will be smaller than 20?

(A) $P\left(z < \dfrac{20-27}{8.5/\sqrt{8}}\right)$

(B) $P\left(z < \dfrac{20-27}{8.5}\right)$

(C) $P\left(z < \dfrac{27-20}{8.5/\sqrt{8}}\right)$

(D) $P\left(z < \dfrac{27-20}{8.5}\right)$

(E) The probability cannot be determined using typical methods since the shape of the sampling distribution is unknown

258. Which of the following statements is not true of the sampling distribution of a sample mean?

(A) The distribution has a mean of $\mu_{\bar{x}} = \mu$ and a standard deviation of $\sigma_{\bar{x}} = \dfrac{\sigma}{\sqrt{n}}$.

(B) It will be normal if the original population is normal.

(C) For small sample sizes, its shape will resemble the shape of the original population.

(D) Its shape will be approximately normal if more than 30 sample means of a given size are found.

(E) It is the distribution of all possible sample means from samples of a given size.

259. A population of unknown shape has a mean of 25 and a standard deviation of 3.4. What is the mean and the standard deviation of the sampling distribution of sample means when $n = 16$?

(A) 25; 0.85

(B) 25; 3.4

(C) 6.25; 0.85

(D) 6.25; 3.4

(E) The mean and the standard deviation cannot be found since the sample size is small

260. A binomial experiment consists of 20 trials. The probability of success on any one trial is 0.37. What is the probability the first success occurs on the fourth trial?

(A) $(0.37)(0.63)^3$

(B) $(0.37)(0.63)^4$

(C) $\dbinom{20}{4}(0.37)^4(0.63)^{16}$

(D) $\dbinom{20}{4}(0.63)^4(0.37)^{16}$

(E) $\dbinom{20}{4}(0.37)^{20}(0.63)^4$

261. Suppose X is a geometric random variable that counts the number of trials necessary to obtain the first successful trial. If the probability of success for any trial is 0.7, what is the probability that the first success will occur on the tenth or eleventh trial?

(A) $\binom{15}{10}(0.7)^{10}(0.3)^5 + \binom{15}{11}(0.7)^{11}(0.3)^4$

(B) $1 - \left[\binom{15}{10}(0.7)^{10}(0.3)^5 + \binom{15}{11}(0.7)^{11}(0.3)^4 \right]$

(C) 0.7

(D) $(0.7)(0.3)^{10} + (0.7)(0.3)^{11}$

(E) $(0.7)(0.3)^9 + (0.7)(0.3)^{10}$

262. Suppose X is a geometric random variable that counts the number of trials necessary to obtain the first successful trial. If the probability of failure for any given trial is 0.42, what is the probability that the first success will occur on the first or second trial?

(A) $(0.58)(0.42)^0 + (0.58)(0.42)^1$

(B) $(0.42)(0.58)^0 + (0.42)(0.58)^1$

(C) $\binom{26}{1}(0.58)^1(0.42)^{25} + \binom{26}{2}(0.58)^2(0.42)^{24}$

(D) $\binom{26}{1}(0.42)^1(0.58)^{25} + \binom{26}{2}(0.42)^2(0.58)^{24}$

(E) $(0.42)(0.58)^1 + (0.42)(0.58)^2$

263. A coin is weighted so that the probability of heads if 0.64. On average, how many coin flips will it take for tails to first appear?

(A) 1.56
(B) 2
(C) 2.78
(D) 6.4
(E) This cannot be determined since the total number of flips is unknown

264. In a binomial experiment with 45 trials, the probability of more than 25 successes can be approximated by $P\left(z > \dfrac{25-27}{3.29}\right)$. What is the probability of success of a single trial of this experiment?

(A) 0.07
(B) 0.56
(C) 0.6
(D) 0.61
(E) 0.79

265. Let X represent the number of successes in 30 trials of a binomial experiment. If $P(X = 2) = \dbinom{30}{2}(0.45)^2(0.55)^{28}$, what is the approximate probability of fewer than 10 successes in the 30 trials?

(A) $P\left(z < \dfrac{30-10}{2.72}\right)$

(B) $P\left(z < \dfrac{2-10}{2.72}\right)$

(C) $P\left(z < \dfrac{10-2}{2.72}\right)$

(D) $P\left(z < \dfrac{13.5-10}{2.72}\right)$

(E) $P\left(z < \dfrac{10-13.5}{2.72}\right)$

266. The distribution of the binomial random variable that counts the number of successful trials in 65, each of which have a 30% probability of success, can be approximated by which of the following distributions?

(A) A slightly skewed left distribution with a mean of 65 and a standard deviation of 19.5.
(B) A slightly skewed left distribution with a mean of 19.5 and a standard deviation of 3.7.
(C) An approximately normal distribution with a mean of 65 and a standard deviation of 19.5.
(D) An approximately normal distribution with a mean of 19.5 and a standard deviation of 3.7.
(E) The number of trials is not large enough to make a determination.

267. The distribution of the binomial random variable that counts the number of successful trials in 70, each of which have a 40% probability of failure, can be approximated by which of the following distributions?

(A) A slightly skewed right distribution with a mean of 28 and a standard deviation of 16.8.

(B) A slightly skewed right distribution with a mean of 42 and a standard deviation of 4.1.

(C) An approximately normal distribution with a mean of 28 and a standard deviation of 16.8.

(D) An approximately normal distribution with a mean of 42 and a standard deviation of 4.1.

(E) The number of trials is not large enough to make a determination.

268. The probability that any one trial out of 23 independent trials will be successful is 17%. Find the probability that fewer than three trials will be successful.

(A) 0.43

(B) 0.15

(C) 0.22

(D) 0.56

(E) 0.78

269. An experiment consists of 31 independent trials. The probability that any one trial is a failure is 45%. Which is the best estimate of the probability that more than 20 trials will be failures?

(A) 0.01

(B) 0.02

(C) 0.11

(D) 0.89

(E) 0.99

270. In a large population, it is known that 55% of items in the population have a certain attribute. Whether an item has the attribute or not is independent of whether any other item has the attribute. If a sample of 95 is drawn from this population, what is the probability that more than 45% of the items in the sample will have the attribute?

(A) $P\left(Z > \dfrac{0.55 - 0.45}{0.051}\right)$

(B) $P\left(Z > \dfrac{0.45 - 0.55}{0.051}\right)$

(C) $P\left(Z > \dfrac{0.55 - 0.45}{9.33}\right)$

(D) $P\left(Z > \dfrac{0.45 - 0.55}{9.33}\right)$

(E) $P\left(Z > \dfrac{42.75 - 52.25}{9.33}\right)$

271. Suppose that X is a binomial random variable that counts the number of successes in 30 trials. Find $P(X > 1)$ if the probability of success for a single trial is 0.14.

(A) 0.0108
(B) 0.0529
(C) 0.0638
(D) 0.9362
(E) 0.9892

272. Suppose that X is a binomial random variable that counts the number of successes in 12 trials. Find $P(X \le 3)$ if the probability of success for a single trial is 0.29.

(A) 0.1807
(B) 0.2460
(C) 0.2775
(D) 0.4765
(E) 0.5235

273. Let X represent the number of successes in n independent trials. If the probability of success for any given trial is p, which of the following is true?

    I. The mean number of successes in n trials is $\sqrt{np(1-p)}$.

    II. The probability of failure for any given trial is $1-p$.

    III. The probability that there will be exactly two successful trials is $\dfrac{2}{n}$.

    (A) I only

    (B) II only

    (C) III only

    (D) I and II only

    (E) I, II, and III

274. The random variable X represents the number of successes in 10 independent trials. Which of the following represents the probability of fewer than two failures?

    (A) $P(X \le 2)$

    (B) $P(X < 2)$

    (C) $P(X > 8)$

    (D) $P(X \ge 8)$

    (E) The probability of success for an individual trial must be known to determine which of these represents the given probability

275. A binomial random variable X counts the number of successes in three trials. Let the probability of success for any given trial be 0.4.

    (a) Find the probability distribution of X in table form.

    (b) Explain how the probability of two or more successes can be found using your table from (a).

276. A small candy company specializes in creating various candies, all of which contain peanuts. Their latest product is the "Chocolate and Peanut Bonanza Bag," a 1-pound (16-ounce) bag of chocolate peanut clusters. The company claims that each bag contains an average of 10 ounces of peanuts and the distribution of the weight of peanuts in this product is approximately normal.

    (a) Suppose that the standard deviation of the weight of peanuts contained in the new product is 1.1 ounces. If the company's claim is true, what is the probability that a sample of 40 Chocolate and Peanut Bonanza bags will contain an average of less than 9.5 ounces of peanuts?

    (b) If the distribution of the weights of peanuts in the new product was not normal but had the same mean and standard deviation, would your calculations in part (a) still be appropriate? Explain.

277. A large population has a skewed right distribution with a mean of 46.1 and a standard deviation of 5.3.

(a) Describe the shape of the sampling distribution of the mean if all possible samples of size 10 are taken.

(b) Describe the shape of the sampling distribution of the mean if all possible samples of size 50 are taken.

(c) Will the mean and the standard deviation be the same in (a) and (b)? Is this true in general or only for this particular problem? Explain.

278. Previous studies have suggested that the average weight of a particular animal species is 121.3 pounds with a standard deviation of 20.9 pounds. The distribution of the weights has not been determined.

(a) A sample of 90 of these animals had an average weight of 115 pounds. What is the probability that a sample of this size would have a mean of 115 pounds or fewer?

(b) Based on your answer in (a), is it likely that the true mean weight of these animals is 121.3 pounds as has been suggested? Justify your answer.

279. A survey of students at a large state university found that 82% had purchased textbooks from an off-campus vendor at least once during their college career.

(a) If 45 students are sampled, how many students would you expect to have purchased textbooks from an off-campus vendor at least once during their college career?

(b) If 150 students are sampled, what is the probability that more than 130 of them have purchased textbooks from an off-campus vendor at least once during their college career?

280. Quality control technicians at a large manufacturing firm have found that approximately 1.2% of parts from the main production line are defective. It is believed that the defects are independent of one another.

(a) An outside consultant is brought in to test parts from the main production line. In a random sample of 150 parts, he finds that four are defective. What is the probability that the consultant would get a sample of this size with four or more defects if the quality control technicians' estimate is correct? Justify your answer.

(b) Assuming the original estimate of 1.2% was correct, if the consultant decides to sample another 150 parts and the first 145 are not defective, what is the probability the 146th part will be defective? Justify your answer.

CHAPTER 7

Confidence Intervals

281. You want to create a 95% confidence interval with a margin of error of no more than 0.05 for a population proportion. The historical data indicate that the population has remained constant at about 0.55. What is the minimum size random sample you need to construct this interval?

(A) 378
(B) 380
(C) 223
(D) 324
(E) None of the above

282. A random sample of 30 households was selected as part of a study on electricity usage, and the number of kilowatt-hours (kWh) was recorded for each household in the sample for the March quarter of 2011. The average usage was found to be 450 kWh. In a very large study in the March quarter of the previous year it was found that the standard deviation of the usage was 81 kWh. Assuming the standard deviation is unchanged and that the usage is normally distributed, provide an expression for calculating a 99% confidence interval for the mean usage in the March quarter of 2011.

(A) $450 \pm 2.756 \times \dfrac{81}{\sqrt{30}}$

(B) $450 \pm 2.575 \times \dfrac{9}{\sqrt{30}}$

(C) $450 \pm 2.33 \times \dfrac{81}{\sqrt{30}}$

(D) $450 \pm 2.575 \times \dfrac{81}{\sqrt{30}}$

(E) None of the above

283. If our sample size from the previous question were tripled, how would this change the confidence interval size?

(A) Divides the interval size by 3

(B) Multiplies the interval size by 1.732

(C) Divides the interval size by 1.732

(D) Triples the interval size

(E) None of the above

284. Other things being equal, which of the following actions will reduce the power of a hypothesis test?

    I. Increasing sample size

   II. Increasing significance level

  III. Increasing beta, the probability of a Type II error

(A) I only

(B) II only

(C) III only

(D) All of the above

(E) None of the above

285. While conducting an experiment to test a hypothesis, we found that our sample size tripled. Which of the following will increase?

    I. The power of the hypothesis test

   II. The effect size of the hypothesis test

  III. The probability of making a Type II error

(A) I only

(B) II only

(C) III only

(D) All of the above

(E) None of the above

286. What is the critical t-value of a 95% confidence interval estimate for a sample size of 25?

(A) 2.06

(B) 1.711

(C) 2.064

(D) 1.708

(E) None of the above

287. Suppose you construct a 95% confidence interval from a random sample of size $n = 20$ with a sample mean 100 taken from a population with unknown μ and a known $\sigma = 10$, and the interval is fairly wide. Which of the following conditions would not lead to a narrower confidence interval?

(A) If you decreased your confidence interval
(B) If you increased your sample size
(C) If the sample size was smaller
(D) If the population standard deviation was smaller
(E) All of the above would lead to a narrower confidence interval

288. Consider the following data:
    1, 7, 3, 3, 6, 4
Assuming these data are drawn from a normal population with mean $= \mu$ and variance $= 6$, the 95% confidence interval for the appropriate z-score is _____ standard units long.

(A) 5.14
(B) 4.9
(C) 4.04
(D) 3.92
(E) None of the above

289. In a middle school with 500 students, it was found that the true proportion of brunettes is 17.8%. If you were to construct a confidence interval estimate of the proportion of brunettes with sample size $n = 50$, which of the following statements would be true?
    I. The interval contains 17.8%.
    II. A 95% confidence interval estimate contains 17.8%.
    III. The center of the interval is 17.8%.

(A) I and II
(B) II and III
(C) I and III
(D) All of the above
(E) None of the above

290. A random sample of the actual weight of 5-lb bags of mulch produces a mean of 4.963 lb and a standard deviation of 0.067 lb. If $n = 50$, which of the following will give a 95% confidence interval for the mean weight (in pounds) of the mulch produced by this company?

(A) 4.963 ± 0.016
(B) 4.963 ± 0.019
(C) 4.963 ± 0.067
(D) 4.963 ± 0.009
(E) None of the above

291. A confidence interval estimate is determined from the SAT scores of an SRS of n students. All other things being equal, which of the following will result in a small margin of error?

   I. Smaller standard deviation
  II. Smaller sample size
 III. Smaller confidence interval

(A) I and II
(B) I and III
(C) II and III
(D) I, II, and III
(E) None of the above

292. In a test to check pH level, 49 samples showed its level to be 2.4 and with a standard deviation of 0.35. Find the 90% confidence interval estimate for the mean pH level.

(A) 2.4 ± 0.01
(B) 2.4 ± 0.08
(C) 2.4 ± 0.32
(D) 2.4 ± 0.35
(E) None of the above

293. In a recent online survey of 1,700 adults it was revealed that only 17% felt the Internet was secure. With what degree of confidence can you say that $17\% \pm 2\%$ of adults believe that online security is secure?

(A) 72.9%
(B) 90%
(C) 95%
(D) 97.2%
(E) 98.6%

294. One of the students in the school council election wants to make healthy lunch menus her pitch to gain votes. What size voter sample would she need to obtain a 90% confidence that her support level margin of error is no more than 4%?

(A) 25
(B) 600
(C) 423
(D) 1691
(E) None of the above

295. Biology professor Doug Miller wants to reduce the width of the confidence interval around the proportion of people who don't wash their hands and therefore carry a certain type of bacteria. What can he do to accomplish this?

    (A) Increase the sample size
    (B) Decrease the sample size
    (C) Change the estimate of the true proportion of people who wash their hands
    (D) Increase the confidence interval
    (E) None of the above

296. What is $t_{0.01}(20)$?

    (A) 2.845
    (B) 2.528
    (C) 2.197
    (D) 1.325
    (E) None of the above

297. You want to create a 95% confidence interval of a population proportion with a margin of error of no more than 0.05. How large of a sample would you need?

    (A) 93
    (B) 200
    (C) 325
    (D) 385
    (E) 1093

298. What is the critical value of t for a 99% confidence interval based on a sample size of 30?

    (A) 2.750
    (B) 2.756
    (C) 2.457
    (D) 2.462
    (E) None of the above

299. When comparing a 95% confidence interval with a confidence interval of 99% created from the same data, how will the intervals differ?

    (A) The sample must be known to determine the difference.
    (B) The mean of the sample must be known to determine the difference.
    (C) The 95% interval will be wider than the 99% interval.
    (D) The 95% interval will be narrower than the 99% interval.
    (E) The use of the t-distribution or the z-distribution will determine how the two intervals differ.

300. Which of these is true about standard error?

    (A) It is another way of expressing standard deviation.

    (B) It is an estimator of standard deviation.

    (C) It is another way of expressing the mean.

    (D) It is an estimator of the mean.

    (E) It is an estimator of confidence interval.

301. What is the effect of sample size on the width of a confidence interval?

    (A) The smaller the sample size, the smaller the width of the confidence interval

    (B) The larger the sample size, the smaller the width of the confidence interval

    (C) The smaller the sample size, the larger the width of the confidence interval

    (D) The larger the sample size, the larger the width of the confidence interval

    (E) None of the above

302. In a particular neighborhood a survey was conducted to determine the percentage of people who did their own gardening. A random sample of 100 householders showed that 63 of them did gardening themselves. An actual 90% confidence interval for householders doing their own gardening is:

    (A) $0.63 \pm 1.960\sqrt{\dfrac{(0.63)(0.37)}{100}}$

    (B) $0.63 \pm 2.560\sqrt{\dfrac{(0.63)(0.37)}{100}}$

    (C) $0.63 \pm 1.645\sqrt{\dfrac{(0.63)(0.37)}{100}}$

    (D) $0.63 \pm 1.960\sqrt{\dfrac{(0.63)(0.37)}{63}}$

    (E) $0.63 \pm 1.645\sqrt{\dfrac{(0.63)(0.37)}{63}}$

303. A recent election poll found that 79% of the population favored the president. If the president's opponents believed this percentage was lower, the null and the alternative hypothesis would be:

(A) *Ho*: $p = 0.79$; *Ha*: $p < 0.79$
(B) *Ho*: $p = 0.79$; *Ha* $p > 0.79$
(C) *Ho*: $p = 0.21$; *Ha*: $p < 0.79$
(D) *Ho*: $p = 0.79$; *Ha*: $p < 0.21$
(E) None of the above

304. In a recent health survey it was found that the average beef consumption in the U.S. per person was 90 pounds per year. A research firm wishes to find whether beef consumption has decreased and does a study to test the null and the alternative hypothesis. The test statistic is $t = -2.147$ with 30 degrees of freedom. Compute the *p*-value for the test.

(A) 0.98
(B) 0.02
(C) 2.147
(D) 0.2
(E) None of the above

305. If the confidence interval for a population proportion changes from 90% to 98% with all else being equal, then _____.

(A) the size of the interval increases by 41%
(B) the size of the interval decreases by 9%
(C) the size of the interval increases by 9%
(D) the size of the interval decreases by 41%
(E) None of the above

306. Which of the following margins of error in a confidence interval results using *z*-scores?

  I. In obtaining the sample surveys errors occurred due to no response
 II. Random sample surveys
III. Errors occurred due to using sample standard deviations as estimates for population standard deviations

(A) I only
(B) II only
(C) III only
(D) I and II
(E) I and III

307. A gallon of gas in 30 test automobiles resulted in $\bar{x} = 28$ and $s = 1.2$. Determine a 95% confidence interval to estimate the mean mileage.

(A) $28 \pm 2.045\left(\dfrac{1.2}{\sqrt{30}}\right)$

(B) $28 \pm 1.960\left(\dfrac{1.2}{\sqrt{30}}\right)$

(C) $28 \pm 2.326\left(\dfrac{1.2}{\sqrt{30}}\right)$

(D) $28 \pm 2.042\left(\dfrac{1.2}{\sqrt{30}}\right)$

(E) $28 \pm 2.750\left(\dfrac{1.2}{\sqrt{30}}\right)$

308. An Australian hospital's administrators wish to learn the average length of stay of all patients. The Australian Bureau of Statistics determines that for a 95% confidence interval the average length of stay needs to be within ± 0.5 days for 50 patients. How many records would need to be looked at to obtain the same confidence interval to be within ± 0.25 days?

(A) 25
(B) 50
(C) 100
(D) 200
(E) 300

309. What is the critical value z of a 99% confidence interval for a known standard deviation?

(A) 1.645
(B) 2.33
(C) 2.345
(D) 2.56
(E) 2.75

310. While weighing five infants in the hospital we found their weights in kilograms to be 3.1, 3.5, 2.8, 3.2, and 3.4. What is a 90% confidence interval estimate of the infants' weights?

(A) 3.2 ± 0.202
(B) 3.2 ± 0.247
(C) 3.2 ± 0.261
(D) 4.0 ± 0.202
(E) 4.0 ± 0.261

311. A poll identifies the proportions of high-income and low-income voters who support a decrease in taxes. If we need to know the answer to be within ± 0.02 at the 95% confidence interval, what sample size should be taken?

(A) 3383
(B) 4803
(C) 7503
(D) 9453
(E) 4503

312. A survey conducted on 1,000 Canadians found that 700 of them refused to receive the H1N1 vaccination. Construct a 95% confidence interval of the estimated proportion of Canadians who refused to receive the vaccination.

(A) (0.723, 0.777)
(B) (0.686, 0.714)
(C) (0.728, 0.672)
(D) (0.986, 0.914)
(E) (0.672, 0.728)

313. For a 95% confidence interval, you want to create a population proportion with a marginal error of no more than 0.025. How large of a sample would you need?

(A) 423
(B) 1537
(C) 385
(D) 2237
(E) 1082

314. In a random sample of 100 balloons, 30 of them are blue. Construct a 95% confidence interval for the true proportion of the blue balloons.

(A) (0.210, 0.399)
(B) (0.193, 0.407)
(C) (0.165, 0.334)
(D) (0.200, 0.380)
(E) (0.198, 0.378)

315. What is the meaning of null hypothesis?

(A) Population parameter
(B) Population mean
(C) Population proportion
(D) Population analogy
(E) Population interval

316. Which of the following statements sounds like typical null or alternative hypotheses?

   I. The coin is fair.

  II. There is no correlation in the population.

 III. The defendant is guilty.

  (A) I only

  (B) I and II

  (C) I and III

  (D) II and III

  (E) I, II, III

317. If a 90% confidence interval is to be constructed, it will be _____ the 95% confidence interval.

  (A) narrower than

  (B) thinner than

  (C) wider than

  (D) the same as

  (E) None of the above

318. A consumer survey found that more than 12% of the screws from a supplier are defective. To test this claim, the correct alternative hypothesis is:

  (A) $p = 0.12$

  (B) $p \neq 0.12$

  (C) $p < 0.12$

  (D) $p > 0.12$

  (E) $p < 0.88$

319. Which of the following statements is true?

   I. If the margin of error is small, then the confidence level is high.

  II. A confidence interval is a type of point estimate.

 III. A population mean is a point estimate.

 IV. If the margin of error is small, then the confidence level is low.

  (A) I only

  (B) II only

  (C) III only

  (D) IV only

  (E) None of the above

320. While conducting an experiment to test a hypothesis, a researcher decides to double her sample size. Which of the following will increase?

   I. The effective size of the hypothesis test

  II. Type II error probability

 III. Power of the hypothesis test

   (A) I only

   (B) II only

   (C) III only

   (D) All of the above

   (E) None of the above

321. A 99% confidence interval for the true mean of a normal population was found to be (123.33, 140.67). If the sample size was 40, find the margin of error for this estimation.

   (A) 8.67

   (B) 17.34

   (C) 264

   (D) 132

   (E) The sample mean must be known to find the margin of error

322. In a sample of 145 cups of coffee, customers rated 37 of them as being served "too hot." Calculate a 95% confidence interval for the true proportion of cups of coffee that customers will feel are served "too hot."

   (A) 36.8% to 37.2%

   (B) 23.5% to 27.5%

   (C) 18% to 33%

   (D) 16% to 35%

   (E) Since the population is not stated to be normal, we cannot use typical techniques to find this interval

323. The 95% confidence interval for a population mean is (12.9, 18.6). Which of the following is the sample mean used to calculate this interval?

   (A) 2.85

   (B) 5.7

   (C) 12.9

   (D) 15.75

   (E) 18.6

324. If all other values are held constant, which of the following is true of a confidence interval if the sample size is increased?

(A) The margin of error will decrease.
(B) The sample mean will increase.
(C) The margin of error will increase.
(D) The population mean will increase.
(E) The variability of the sample will decrease.

325. A doctor wishes to survey a sample of patients to determine what percentage are happy with the recent changes to the appointment system. If the doctor wishes the estimate to be within two percentage points of the true percentage at a 95% confidence interval, how many patients should be surveyed?

326. Out of a random sample of 1,000 kids asked whether they had ever downloaded applications for their handheld gadgets, 750 of them responded "yes."

(a) Construct and interpret a 99% confidence interval for the population p, the true proportion of the population who have ever downloaded applications for their handheld gadgets.

(b) How large would the sample need to be to estimate p within 2% at a 99% confidence level?

327. A local high school baseball team played 81 games at home and 81 games away. The team won 48 of their home games and 44 of the games played away. We can consider these games as samples from a potentially large number of games played at home and away.

(a) Identify the populations and parameters of interest.

(b) Construct and interpret a 90% confidence interval for the difference between the proportion of the games that the team wins at home and the games they win when they're away.

328. In a sample of 10 baseball players the mean income was $800,000 with a standard deviation of $315,000.

(a) Assuming all necessary conditions are met, find the 95% confidence interval of the mean salary of the baseball players.

(b) What assumptions are necessary for the above estimate?

329. A manufacturer of televisions receives shipments of LCD panels from an overseas supplier. It is not cost-effective to inspect each of the panels for defects, so a sample is taken from each shipment. A significance test is conducted to determine whether the proportion of the defective panels is greater than the acceptable limit of 1%. If it is, the shipment will be returned to the supplier. Essentially, this is a test of H_0: $\pi = 0.01$ versus H_a: $\pi > 0.01$, where p is the true proportion of defective panels in the shipment.

(a) Describe the Type I and Type II errors in the context of the situation.
(b) If you were the supplier of the LCD panels, which error would be more serious and why?
(c) If you were the TV manufacturer, which error would be more serious and why?

330. The table below shows data from Canadian and U.S. ninth-grader test scores.

	Sample Size	Sample Mean	Sample Standard deviation
U.S.	300	95	4.5
Canadians	250	90.5	4.1

(a) Write the formula for the 95% confidence interval in terms of the values shown above.
(b) Determine the confidence interval.
(c) Provide an interpretation of the confidence interval. Include in your interpretation a conclusion that can be drawn about which country has a higher average test score.

Inference for Means and Proportions

331. A scientist is testing whether a new fertilizer increases the height of a particular plant. As part of the experiment, 15 of the plants are randomly assigned the new fertilizer and 15 are randomly assigned an old formula. What type of test should be used to determine whether the average height of the plants is higher using the new fertilizer?

(A) A matched-pairs t-test

(B) A one-sample proportion z-test

(C) A two-sample t-test

(D) A two-sample proportion z-test

(E) A chi-squared test of association

332. In a survey of 550 students this year, 260 stated that they had worked full time during their summer break. In the same survey last year, 300 of 610 students stated the same. Which type of test should be used to determine if the percentage of students who worked full time is higher this year than last year?

(A) A matched-pairs t-test

(B) A one-sample proportion z-test

(C) A two-sample t-test

(D) A two-sample proportion z-test

(E) A chi-squared test of association

333. The mean weight of 120 male high school seniors is 156 pounds with a standard deviation of 23 pounds. Assuming the distribution of weights is approximately normal, which type of test should be used to determine whether the mean weight of all high school seniors is more than 150 pounds?

(A) A matched-pairs t-test

(B) A one-sample proportion z-test

(C) A two-sample t-test

(D) A one-sample t-test

(E) A two-sample proportion z-test

334. In 1999, 45% of freshman at a particular college had enrolled in an introductory English course. In a sample of 800 students this year, 460 freshman had enrolled in an introductory English course. Which type of test should be used to determine whether the percentage has increased since 1999?

(A) A matched-pairs t-test
(B) A one-sample proportion z-test
(C) A two-sample t-test
(D) A one-sample t-test
(E) A two-sample proportion z-test

335. Two years ago, it was determined that the average age of viewers of an evening television program was 56.4 years. A study of 50 randomly selected viewers this year found that the average age was 55.1 years. Which type of test should be used to determine whether the average age has decreased?

(A) A matched-pairs t-test
(B) A one-sample proportion z-test
(C) A two-sample t-test
(D) A one-sample t-test
(E) A two-sample proportion z-test

336. A sample of 50 employees with only a high school education had an average yearly salary of $32,000 while a sample of 45 employees with a bachelor's degree had an average yearly salary of $41,000. Which type of test should be used to determine whether the average salary is higher for employees with a bachelor's degree than with only a high school education?

(A) A matched-pairs t-test
(B) A one-sample proportion z-test
(C) A two-sample t-test
(D) A one-sample t-test
(E) A two-sample proportion z-test

337. In a test of $H_0 : \mu \geq 12.5$ versus $H_a : \mu < 12.5$, the null hypothesis is not rejected at the 5% level. Which of the following statements must be true?
  I. In the same test, the null hypothesis will be rejected at the 10% level.
  II. The sample used had a mean that was 12.5.
  III. There is no evidence at the 5% level that the true mean is less than 12.5.

(A) I only
(B) II only
(C) III only
(D) I and II only
(E) I, II, and III

338. Which of the following statements must be true to use a one-sample proportion z-test?

I. $np \geq 5$, $n(p - 1) \geq 5$

II. The population the sample is taken from is approximately normal.

III. The hypothesized proportion is greater than 30%.

(A) I only
(B) II only
(C) III only
(D) I and III only
(E) I, II, and III

339. If a two-sided hypothesis test of the mean rejects the null hypothesis at the 10% level, which of the following statements must be true of a 90% confidence interval calculated on the same data?

(A) The confidence interval will contain the hypothesized value of the mean.
(B) The confidence interval will not contain the hypothesized value of the mean.
(C) The confidence interval will contain values within 10% of the hypothesized mean.
(D) The confidence interval will only contain values larger than the hypothesized mean.
(E) The confidence interval will only contain values smaller than the hypothesized mean.

340. A test of $H_0 : \mu = 7$ versus $H_a : \mu \neq 7$ does not reject the null hypothesis at the 5% level. If calculated using the same sample data, which of the following is a possible 95% confidence interval for the population mean?

(A) 8.4 ± 1.1
(B) 9.2 ± 3.4
(C) 6.2 ± 0.5
(D) 8.4 ± 0.9
(E) There is not enough information to determine anything about a 95% confidence interval

341. The average monthly cell phone bill from Company A for a sample of 45 bills was \$49.00 with a standard deviation of \$10.50, while the average monthly cell phone bill from Company B for a sample of 45 bills was \$52.00 with a standard deviation of \$12.10. Which of the following is the appropriate test statistic to test whether the average bill from Company A is lower than Company B?

(A) $\dfrac{49-52}{\sqrt{\dfrac{10.5^2}{45}+\dfrac{12.1^2}{45}}}$

(B) $\dfrac{49}{\dfrac{10.5}{\sqrt{45}}}$

(C) $\dfrac{52}{\dfrac{12.1}{\sqrt{45}}}$

(D) $\dfrac{3}{\dfrac{11.5}{\sqrt{45}}}$

(E) $\dfrac{49-52}{\sqrt{11.5^2\left(\dfrac{1}{45}+\dfrac{1}{45}\right)}}$

342. A random sample of 20 from a normal population has a mean of 15.1. In a test of $H_0 : \mu = 15$ versus $H_a : \mu \neq 15$, which of the following is the correct decision and conclusion at the 5% level?

(A) Reject H_0 concluding that there is evidence the true mean is 15.
(B) Reject H_0 concluding that there is no evidence the true mean is 15.
(C) Do not reject H_0 concluding that there is evidence the true mean is 15.
(D) Do not reject H_0 concluding that there is no evidence that the true mean is 15.
(E) There is not enough information to determine whether the null hypothesis should be rejected.

343. In a test of $H_0 : p \geq 0.5$ versus $H_a : p = 0.5$, the point estimate of the population proportion is 0.51. Which of the following is the correct conclusion?

(A) At the 5% level there is evidence that the population proportion is less than 0.5.

(B) At the 5% level there is no evidence that the population proportion is less than 0.5.

(C) At the 5% level there is evidence that the population proportion is less than 0.5, but not at the 10% level.

(D) At the 5% level there is no evidence that the population proportion is less than 0.5, but there is evidence at the 10% level.

(E) A conclusion cannot be made based on this point estimate alone.

344. A sample of 15 from a normal population yields a sample mean of 43 and a sample standard deviation of 4.7. What is the p-value that should be used to test the claim that the population mean is less than 45?

(A) 0.0608

(B) 0.1216

(C) 0.4696

(D) 0.9392

(E) The p-value cannot be determined from the given information

345. A popular website claims to receive an average of 4,500 unique visitors a day. An interested advertiser takes a random sample of 20 days and finds a sample average for the number of unique visitors in a day. Based on a further analysis of the data, it appears reasonable to assume the population is approximately normal and a t-test statistic is calculated to test whether the true population mean is 4,500. If the t-test statistic is -1.73, what is the approximate p-value of this test?

(A) $0.0125 < p < 0.025$

(B) $0.025 < p < 0.05$

(C) $0.05 < p < 0.1$

(D) $0.1 < p < 0.2$

(E) The approximate p-value cannot be determined without the sample mean

346. A 99% confidence interval for the mean of a normal population is 153.7 ± 12.1. In a test of $H_0 : \mu = 156$ versus $H_a : \mu \neq 156$, what would be the appropriate decision and conclusion at the 1% level?

(A) Reject H_0 concluding that there is evidence the true mean is not 156.

(B) Reject H_0 concluding that there is no evidence the true mean is not 156.

(C) Do not reject H_0 concluding that there is evidence the true mean is not 156.

(D) Do not reject H_0 concluding that there is no evidence that the true mean is not 156.

(E) H_0 may or may not be rejected depending on the population standard deviation.

347. To estimate the percentage of employees who approve of a new policy, a company takes a sample and calculates a 95% confidence interval for the percentage of all employees who approve of a new policy. It is believed that the percentage is approximately 30%. If the calculated confidence interval is (0.345, 0.511), is there evidence to support this belief at the 5% level?

(A) Yes, since the confidence interval estimate is only 95% accurate.

(B) No, since the confidence interval does not contain 0.3.

(C) Yes, since 0.345 is within five percentage points of 0.3.

(D) No, since the true percentage of all employees who approve must be between 34.5% and 51.1%.

(E) It is not possible to determine whether there is evidence without the test statistic p-value.

348. A poll is conducted to determine whether more than half the pedestrians at a particular intersection think a crosswalk should be installed. If 552 of 759 sampled pedestrians think a crosswalk should be installed, what null and alternative hypothesis should be used to determine whether more than half of all pedestrians at this intersection believe this?

(A) $H_0 : p > 0.5$ versus $H_a : p < 0.5$

(B) $H_0 : p = 0.5$ versus $H_a : p \neq 0.5$

(C) $H_0 : p \leq 0.5$ versus $H_a : p > 0.5$

(D) $H_0 : p \geq 552$ versus $H_a : p < 552$

(E) $H_0 : p \leq 552$ versus $H_a : p > 552$

349. A sample of 45 random vehicles traded in at a particular car dealership had an average odometer reading of 35,110 miles with a standard deviation of 6,235 miles. If the manager wishes to determine whether there is evidence that the mean odometer reading of all cars traded in is more than 35,000 miles, which null and alternative hypothesis should he use?

(A) $H_0 : \mu > 35,110$ versus $H_a : \mu < 35,110$
(B) $H_0 : \mu = 35,000$ versus $H_a : p \neq 35,000$
(C) $H_0 : \mu \leq 35,000$ versus $H_a : \mu > 35,110$
(D) $H_0 : p \geq 35,110$ versus $H_a : p < 35,110$
(E) $H_0 : \mu \leq 35,000$ versus $H_a : \mu > 35,000$

350. Which of the following is true of a paired samples t-test?
  I. The point estimate used is the average of the differences.
  II. There may be one, two, or three samples.
  III. The standard normal distribution is used to find the p-value.

(A) I only
(B) II only
(C) III only
(D) I and II only
(E) I, II, and III

351. A test of $H_0 : \mu = 0$ versus $H_a : \mu \neq 0$ is performed using a sample of 20 from an approximately normal population. If the sample standard deviation was 2.16 and the t-test statistic was 1.249, find the mean of the sample.

(A) 0
(B) 0.3867
(C) 0.6033
(D) 7.734
(E) 9.6598

352. In a sample of 390 people attending a concert, 60% stated that the venue was a good location for that type of concert. What test statistic should be used to test $H_0 : p \leq 0.65$ versus $H_a : p > 0.65$?

(A) −2.07
(B) −2.067
(C) 0
(D) 2.067
(E) 2.07

353. The null hypothesis $H_0 : p \leq 0.6$ was rejected at the 10% level but not at the 5% level. Which of the following statements must be true?

(A) The sample proportion was within 5–10% of 0.6.
(B) The p-value of the test is less than 0.1 but more than 0.05.
(C) The sample proportion is less than 0.1 but more than 0.05.
(D) The p-value of the test is greater than 0.1.
(E) The sample proportion is greater than 0.1.

354. A researcher wishes to test whether two population means are different. If he uses two dependent samples, which of the following would be the correct alternative hypothesis?

(A) $\mu_1 = \mu_2$
(B) $\mu_1 \neq \mu_2$
(C) $\mu_d = 0$
(D) $\mu_d \neq 0$
(E) Either (B) or (D) could be used as the alternative hypothesis

355. In which of the following cases is a pooled estimate of the population standard deviation used?

(A) When performing a two-sample t-test and the population variances are believed to be the same.
(B) When performing a two-sample t-test and the population variances are believed to be different.
(C) When performing a two-sample t-test and the population variances are unknown.
(D) When performing a two-sample t-test regardless of the relationship between the population variances.
(E) When performing a two-sample t-test and the population variances are known.

356. The p-value of a test of $H_0 : \mu_1 \leq \mu_2$ versus $H_a : \mu_1 > \mu_2$ is 0.04711. Determine the p-value of the test if the alternative was $H_a : \mu_1 \neq \mu_2$ instead.

(A) 0.0236
(B) 0.0471
(C) 0.0942
(D) 0.9058
(E) 0.9529

357. The p-value for a test of $H_0 : p \geq 0.25$ versus $H_a : p < 0.25$ is 0.0625. Which of the following would be the correct p-value for a test of $H_0 : p \leq 0.25$ versus $H_a : p > 0.25$ if the same sample data are used?

(A) 0.0313
(B) 0.0625
(C) 0.125
(D) 0.9375
(E) The number of successes in the sample and the sample size need to be known to determine the p-value

358. In a sample of 500 students from statistics courses in the fall semester, 312 earned a grade of C or higher. In the spring semester, 400 of the 525 students sampled earned a grade of C or higher. If we treat the fall semester as "sample 1" and the spring semester as "sample 2," what is the appropriate standardized test statistic to determine whether a smaller proportion of students passed in the fall?

(A) $z = -6.699$
(B) $z = -4.792$
(C) $z = 4.792$
(D) $z = 6.699$
(E) The z-test statistic depends on the level of significance of the test

359. A firm is hired by a small town to estimate the proportion of large trucks that are carrying hazardous materials on the main highway through the town. Of 350 randomly selected trucks, 92 were carrying hazardous materials. Is there evidence at the 10% level that the proportion of large trucks carrying hazardous materials is less than 30%?

(A) Yes, since the sample proportion is less than 30%.
(B) No, since the sample proportion is larger than 10%.
(C) Yes, since the p-value of the one-sided test is less than 10%.
(D) No, since the p-value of the one-sided test is less than 10%.
(E) The sample size is too small to make a determination.

360. The p-value for a one-sided test of the population mean is 0.0513. Which of the following would be true if the sample size is increased but all other sample statistics remain the same?

(A) The absolute value of the standardized test statistic will be larger while the p-value will be smaller.

(B) Both the absolute value of the standardized test statistic and the p-value will be larger.

(C) The absolute value of the standardized test statistic will be smaller while the p-value will be larger.

(D) Both the absolute value of the standardized test statistic and the p-value will be smaller.

(E) The p-value will increase or decrease by a factor of n where n is the sample size.

361. A sample of 14 from an approximately normal population is used to test $H_0 : \mu \leq 5$ versus $H_a : \mu > 5$. If the p-value of this test is 0.0329, which of the following is the correct decision and conclusion at the 5% level?

(A) Reject the null hypothesis concluding that there is evidence the true mean is larger than 5.

(B) Do not reject the null hypothesis concluding that there is no evidence the true mean is larger than 5.

(C) Reject the null hypothesis concluding that there is no evidence the true mean is larger than 5.

(D) Do not reject the null hypothesis concluding that there is evidence the true mean is larger than 5.

(E) There is not enough information to determine whether the null hypothesis should be rejected.

362. A sample of 20 from an approximately normal population is used to test $H_0 : \mu \geq 12$ versus $H_a : \mu < 12$. The t-test statistic for this test was determined to be -1.002. Which of the following is the correct conclusion at the 1% level?

(A) Reject the null hypothesis concluding that there is evidence the true mean is less than 12.

(B) Do not reject the null hypothesis concluding that there is no evidence the true mean is less than 12.

(C) Reject the null hypothesis concluding that there is no evidence the true mean is less than 12.

(D) Do not reject the null hypothesis concluding that there is evidence the true mean is smaller than 12.

(E) There is not enough information to determine whether the null hypothesis should be rejected.

363. The p-value for a test of $H_0 : p \geq 0.52$ versus $H_a : p < 0.52$ is 0.4872. Which of the following is the correct conclusion at the 5% level?

(A) Reject the null hypothesis concluding that there is evidence the true proportion is less than 0.52.

(B) Do not reject the null hypothesis concluding that there is no evidence the true proportion is less than 0.52.

(C) Reject the null hypothesis concluding that there is no evidence the true proportion is less than 0.52.

(D) Do not reject the null hypothesis concluding that there is evidence the true proportion is less than 0.52.

(E) Since 0.4872 is less than 0.52, it is not necessary to perform a hypothesis test to determine that the true proportion is less than 0.52.

364. The z-test statistic for a test of $H_0 : p \leq 0.31$ versus $H_a : p > 0.31$ is 2.325. Which of the following is the correct conclusion at the 5% level?

(A) Reject the null hypothesis concluding that there is evidence the true proportion is greater than 0.31.

(B) Do not reject the null hypothesis concluding that there is no evidence the true proportion is greater than 0.31.

(C) Reject the null hypothesis concluding that there is no evidence the true proportion is greater than 0.31.

(D) Do not reject the null hypothesis concluding that there is evidence the true proportion is greater than 0.31.

(E) The point estimate of the proportion must be known to determine whether to reject the null hypothesis.

365. Two independent samples are taken from two approximately normal populations to test $H_0 : \mu_1 \geq \mu_2$ versus $H_a : \mu_1 < \mu_2$. If the t-test statistic is -1.209, which of the following is the correct conclusion at the 10% level?

(A) Do not reject the null hypothesis and conclude that there is evidence the true mean of the first population is smaller than that of the second.

(B) Do not reject the null hypothesis and conclude that there is no evidence the true mean of the first population is smaller than that of the second.

(C) Reject the null hypothesis and conclude that there is evidence the true mean of the first population is smaller than that of the second.

(D) Reject the null hypothesis and conclude that there is no evidence the true mean of the first population is smaller than that of the second.

(E) The sizes of each sample must be known in order to determine whether to reject the null hypothesis.

366. Two independent samples are taken from two approximately normal populations to test $H_0 : \mu_1 \leq \mu_2$ versus $H_a : \mu_1 > \mu_2$. If the p-value is determined to be 0.0223, which of the following is the correct conclusion at the 10% level?

(A) Do not reject the null hypothesis and conclude that there is evidence the true mean of the first population is larger than that of the second.

(B) Do not reject the null hypothesis and conclude that there is no evidence the true mean of the first population is larger than that of the second.

(C) Reject the null hypothesis and conclude that there is evidence the true mean of the first population is larger than that of the second.

(D) Reject the null hypothesis and conclude that there is no evidence the true mean of the first population is larger than that of the second.

(E) The sizes of each sample must be known in order to determine whether to reject the null hypothesis.

367. A sample of seven randomly selected drivers are given a test to determine their reaction time to a sudden change in the driving course both with and without a front seat passenger. If the p-value to determine whether the presence of a passenger increases reaction time is 0.0849, which of the following is the correct conclusion at the 5% level?

(A) Using a two-sample t-test there is no evidence that the reaction time is increased.

(B) Using a two-sample t-test there is evidence that the reaction time is increased.

(C) Using a paired samples t-test there is no evidence that the reaction time is increased.

(D) Using a paired samples t-test there is evidence that the reaction time is increased.

(E) It is not possible to test whether there has been a change in reaction time using either the two-sample t-test or the paired samples t-test.

368. Sixteen randomly selected high school football players are timed running a 40-yard dash prior to a specially designed running program. The same players are then timed running the 40-yard dash after the program. If the t-test statistic to determine whether the training program increases the speed of those running the 40-yard dash is 1.115, which of the following is the correct conclusion at the 1% level?

(A) Using a two-sample t-test there is no evidence that the speed has increased.

(B) Using a two-sample t-test there is evidence that the speed has increased.

(C) Using a paired samples t-test there is no evidence that the speed has increased.

(D) Using a paired samples t-test there is evidence that the speed has increased.

(E) It is not possible to test the statement without the players' average time running the 40-yard dash.

369. A survey of randomly selected visitors to a mall yielded the following data:

	Approve of a New Design Plan for the Seating in the Food Court	Number Surveyed
Male	152	254
Female	199	301

Based on this sample, what would be the correct decision and conclusion of a test to determine whether the proportion who approve of a new design plan for the seating area in the food court is higher for women than it is for men at the 5% level?

(A) Reject the null hypothesis concluding that there is evidence the proportion is higher for women.

(B) Do not reject the null hypothesis concluding that there is no evidence that the proportion is higher for women.

(C) Reject the null hypothesis concluding that there is no evidence that the proportion is higher for women.

(D) Do not reject the null hypothesis concluding that there is no evidence that the proportion is higher for women.

(E) Conclude the proportion is higher for women since 199/301 > 152/254.

370. Which of the following statements is true regarding a two-proportion z-test?
  I. The p-value can be calculated using the standard normal distribution.
  II. The test uses two independent or dependent samples.
  III. The sample size from both populations must be the same.

  (A) I only
  (B) II only
  (C) III only
  (D) I and II only
  (E) I, II, and III

371. A realtor states that the mean sale price for single-family homes in the county is \$290,000. If $\mu =$ the average price of all the houses concerned, the null and alternative hypotheses you would use to test this statement are:

  (A) $H_0 : \mu > 290{,}000; H_A : \mu \leq 290{,}000$
  (B) $H_0 : \mu = 290{,}000; H_A : \mu \neq 290{,}000$
  (C) $H_0 : p = 290{,}000; H_A : p \neq 290{,}000$
  (D) $H_0 : \mu \leq 290{,}000; H_A : \mu > 290{,}000$
  (E) $H_0 : \mu > 290{,}000; H_A : \mu \leq 290{,}000$

372. Which of the following statements is true?
  I. Tests of significance (hypothesis tests) are designed to test the strength of evidence against the null hypothesis.
  II. If the p-value for a test is 0.125, the probability that the alternative hypothesis is true is 0.875.
  III. The alternative hypothesis is two-sided if deviations from the null hypothesis in both directions are being considered.

  (A) I, II, and III
  (B) None of the above
  (C) I and III
  (D) II and III
  (E) I and II

373. Which of the following statements is true?
  I. The null hypothesis is stated in terms of a population parameter.
  II. The alternative hypothesis is stated in terms of a sample statistic.
  III. The larger the p-value the weaker the evidence against the null hypothesis.

  (A) I and II only
  (B) I and III only
  (C) III only
  (D) None of the above
  (E) I, II, and III

374. Which of the following conditions are necessary to justify the use of t-procedures in a significance test when doing inference for a population mean?
   I. The sample is a simple random sample from the population.
   II. The sample size is large (rule of thumb: $n \geq 50$) or the population from which the sample is drawn is approximately normally distributed.
   III. The sample size is large (rule of thumb: $n \geq 30$) or the population from which the sample is drawn is approximately normally distributed.

(A) I and II only
(B) I only
(C) III only
(D) None of the above
(E) I and III only

375. Which of the following statements are true?
   I. z-Procedures are always preferable to t-procedures when doing inference for means.
   II. When considering use of z-procedures in doing inference for population means, make sure the population from which the sample is drawn is normally distributed.
   III. When considering use of z-procedures in doing inference for population means, make sure the population from which the sample is drawn is binomially distributed.

(A) III only
(B) None of the above
(C) I and III only
(D) II only
(E) I and II only

376. If the null hypothesis is given as $H_0 : \mu_1 - \mu_2 = 0$, a one-sided alternative hypothesis would be:
   I. $H_A : \mu_1 - \mu_2 \neq 0$
   II. $H_A : \mu_1 - \mu_2 > 0$
   III. $H_A : \mu_1 - \mu_2 < 0$

(A) I only
(B) II only
(C) None of these are valid alternative hypotheses
(D) II or III only
(E) It cannot be determined from the information given

377. Which of the following statements are true?
- I. When doing inference for a population mean we usually use t-procedures rather than z-procedures.
- II. When doing inference for a population proportion we seldom use z-procedures because the sample size is so large.
- III. t-Procedures are said to be "robust" when they still work reasonably well even if the assumption of normality is violated.

(A) III only
(B) None of the above
(C) I and III only
(D) II only
(E) I and II only

378. Which of the following are true about the null hypothesis $(H_\downarrow(0))$?
- I. H_0 is the hypothesis we are testing.
- II. H_0 can only be rejected when a specific significance level is stated.
- III. H_0 is rejected in error when a Type I error occurs.

(A) III only
(B) I, II, and III
(C) I and III only
(D) I and II only
(E) I only

379. At a dairy bottling plant, a machine is supposed to fill cartons with 64 ounces of milk but the plant manager suspects that the machine is overfilling the cartons. He tests this by sampling 49 cartons of milk. If the mean content of the cartons is greater than 64.1, he will reject the 64-ounce hypothesis. The machine is in fact operating correctly, with a mean fill volume of 64.0 ounces and a standard deviation of 0.5 ounces. What is the probability that the manager will mistakenly reject the 64-ounce claim?

(A) 0.09
(B) 0.11
(C) 0.081
(D) 0.153
(E) 0.01

380. A real estate newsletter claims that the mean number of house sales per year for real estate agencies in the southeastern U.S. is 137. A reporter for the real estate blog *SellsFast* suspects that the number is inflated. The reporter tests this by sampling 32 agencies. The sample mean is 131 with a standard deviation of 20. Where is the *p*-value?

(A) Below 0.01
(B) Between 0.01 and 0.025
(C) Between 0.025 and 0.05
(D) Between 0.05 and 0.10
(E) Over 0.10

381. Which of the following are necessary conditions for a paired samples *t*-test?
  I. The sample size is larger than 30 or the differences follow an approximately normal distribution.
  II. A two-sample *t*-test does not yield any results.
  III. The sample is obtained using simple random sampling.

(A) I only
(B) II only
(C) III only
(D) I and II only
(E) I and III only

382. A one-sided test is used to determine whether there is evidence a population mean is larger than 5. If the sample mean was 5.3, which of the following probability statements represent the appropriate *p*-value?

(A) $p(\bar{x} \geq 5.31, \mu = 5)$
(B) $p(\bar{x} \geq 5.31, \mu > 5)$
(C) $p(\bar{x} \leq 5.31, \mu = 5)$
(D) $p(\mu > 5)$
(E) $p(\mu < 5)$

383. In a sample of *n* individuals, *x* individuals in the sample had a characteristic of interest. If *n* is increased while *x* remains the same, the *p*-value for a one-proportion test of $H_0 : p \geq 0.5$ versus $H_a : p < 0.5$ will

(A) increase since the sample proportion will decrease.
(B) decrease since the sample proportion will decrease.
(C) remain the same since the sample size is constant.
(D) either increase or decrease depending on the sample size.
(E) either increase or decrease based on the sampling procedure used.

384. Which of the following statements is true of the null hypothesis in any hypothesis test?
   I. It is rejected only when there is convincing evidence that it is false.
   II. If it is not rejected then it is a true statement.
   III. The probability of rejecting the null hypothesis is the level of significance.

   (A) I only
   (B) II only
   (C) III only
   (D) I and II only
   (E) I and III only

385. In a test of $H_0 : p \leq 0.20$ versus $H_a : p > 0.20$, if the null hypothesis is rejected at the 0.05 level, which of the following statements must be true of the standardized test statistic?

   (A) It is larger than 1.645.
   (B) It is smaller than −1.645.
   (C) It is between −1.645 and 1.645.
   (D) It is either larger than 1.645 or smaller than −1.645.
   (E) It may be any value depending on the sample size used.

386. A small non-profit that assists at-risk children in afterschool programs conducts a study of two different neighborhoods to determine whether there is a difference in the proportion of families with children living below the poverty line. The data collected are below.

	Number of Families with Children Living below the Poverty Line	Number of Families Surveyed
Neighborhood 1	258	392
Neighborhood 2	481	517

   (a) What are the point estimates for the proportion of families with children living below the poverty line in each neighborhood?
   (b) Is there evidence of a difference between the two neighborhoods? Justify your answer.

387. As part of the development process for a retailers' informational brochure, a company that owns a mall is studying the average amount of money that customers spend in its stores. A random sample of 127 customers spent an average of $215.67 with a standard deviation of $81.90. To determine whether it is appropriate to make the statement, "Our customers spend an average of more than $200 per visit," a hypothesis test is performed and a p-value of 0.0165 is found.

    (a) State the appropriate null and alternative hypotheses to test the quoted statement.

    (b) Explain exactly what the p-value represents in this situation.

388. Suppose that the data below were collected from a sample of gas stations in Indiana.

Number of Stations Selling E85	Number of Sampled Stations
251	412

    (a) Calculate a 95% confidence interval for the percentage of all gas stations in Indiana selling E85.

    (b) Using your confidence interval only, would the null hypothesis $H_0 : p = 0.61$ be rejected at the 5% level? Justify your answer.

389. A vitamin supplement is purported to increase the short-term memory of those who take it. To determine whether this statement is valid, researchers had a sample of nine individuals take a memory test both with and without the supplement. The test had the participants recite a list of 50 short words right after hearing the list, and a different list was used for each part of the experiment. The data collected are below.

Participant	1	2	3	4	5	6	7	8	9
Number of words correctly recited without the supplement	21	19	17	26	35	11	8	40	15
Number of words correctly recited with the supplement	22	18	18	41	30	13	10	42	17

    (a) What are some possible factors other than the supplement that may have affected the results of this study?

    (b) Assuming the necessary conditions hold, is there evidence that the supplement increased the number of words correctly recalled? Justify your answer.

390. Suppose that in a test of $H_0 : \mu = 4.5$ versus $H_a : \mu \neq 4.5$, the calculated p-value is 0.558.

    (a)   Determine whether there is evidence at the 0.05 level that the mean is larger than 4.5.

    (b)   Can the given information be used to determine whether the mean is larger than 4.7? Explain.

CHAPTER 9

Inference for Regression

391. In order to understand the relationship between two variables x and y, a sample of 50 pairs of data (x, y) is taken. A test of $H_0 : \beta = 0$ versus $H_0 : \beta \neq 0$ leads to a p-value of 0.0025. Which of the following must be true if the assumptions for the appropriateness of the test are valid?

(A) At the 5% level, there is a linear relationship between the variables.
(B) There is a probability of 0.0025 that there is a linear relationship between the variables.
(C) At the 1% level, the null hypothesis should not be rejected.
(D) There is a strong positive linear correlation between the variables.
(E) A confidence interval for β will not contain zero.

392. A teacher is investigating the relationship between hours spent studying and exam scores (scale 0–100).

Predictor	Coef	St Dev	T	P
Constant	59.2	2.415	24.53	0.000
Hours	1.718	0.22	7.797	0.000

$s = 4.74$ R-Sq $= 77.15\%$ R-Sq(adj) $= 75.88\%$

Based on the computer output above, in a test of $H_0 : \beta = 0$ versus $H_a : \beta \neq 0$, which of the following would be correct?

(A) Reject H_0 concluding that at the 5% level there is a linear relationship between hours spent studying and exam scores.
(B) Do not reject H_0 concluding that at the 5% level there is a linear relationship between hours spent studying and exam scores.
(C) Reject H_0 concluding that at the 5% level there is no evidence of a linear relationship between hours spent studying and exam scores.
(D) Do not reject H_0 concluding that at the 5% level there is no evidence of a linear relationship between hours spent studying and exam scores.
(E) Based on the large value of the constant, conclude there is a linear relationship between hours spent studying and exam scores.

393. Which of the following is an appropriate test to determine whether there is a linear relationship between two variables x and y?

(A) A two-sample t-test of $H_0 : \mu_X = \mu_Y$ versus $H_a : \mu_X \neq \mu_Y$

(B) A two-sample z-test of $H_0 : \mu_X = \mu_Y$ versus $H_a : \mu_X \neq \mu_Y$

(C) A t-test of $H_0 : \beta = c$ versus $H_a : \beta \neq c$ where β is the slope of the regression line and c is any constant

(D) A t-test of $H_0 : \beta_0 = 0$ versus $H_a : \beta_0 > 0$ where β_0 is the y-intercept of the regression line

(E) A t-test of $H_0 : \beta = 0$ versus $H_a : \beta \neq 0$ where β is the slope of the regression line

394. The relationship between two variables x and y is being studied. Using sample data, it is found that the correlation coefficient $r = 0$. Which of the following statements is not true?

(A) There is not a linear relationship present between the variables in the sample.

(B) The y-intercept of the least-squares line is zero.

(C) The slope of the least-squares line is zero.

(D) A confidence interval for the slope of the line in the population will contain zero.

(E) A hypothesis test of $H_0 : \beta = 0$ versus $H_a : \beta \neq 0$ may reject the null hypothesis.

395. Based on a sample of 20 data values, the least-squares regression line to predict y from x is $\hat{y} = 21.34 + 0.09x$. If $S_b = 0.76$, find the t-test statistic for $H_0 : \beta = 0$.

(A) 0.09

(B) 0.12

(C) 0.67

(D) 8.44

(E) 28.08

396. A 95% confidence interval for the slope of a regression line is 0.0935 ± 1.7128. If testing $H_0 : \beta = 0$ versus $H_a : \beta \neq 0$, which of the following must be true?

(A) The null hypothesis will be rejected at the 5% level since the confidence interval contains values larger than 1.

(B) The null hypothesis will not be rejected at the 5% level since the estimated slope is small.

(C) The null hypothesis will not be rejected at the 5% level since the confidence interval contains zero.

(D) It cannot be determined whether the null hypothesis will be rejected since S_b is not given.

(E) It cannot be determined whether the null hypothesis will be rejected since the margin of error is large compared to the estimated slope.

397. A researcher wishes to determine whether there is a positive linear relationship between two variables x and y. If β represents the slope of the regression line, which of the following null and alternative hypotheses should be used?

(A) $H_0 : \beta = 0$ versus $H_a : \beta > 0$
(B) $H_0 : \beta = 0$ versus $H_a : \beta < 0$
(C) $H_0 : \beta > 0$ versus $H_a : \beta < 0$
(D) $H_0 : \beta \geq 0$ versus $H_a : \beta < 0$
(E) $H_0 : \beta < 0$ versus $H_a : \beta \geq 0$

398. A student studying the relationship between the engine temperature of a car at 60 mph and gas mileage at the same speed believes there is a negative linear relationship between the two variables. What null and alternative hypotheses should be used in order to test this belief?

(A) $H_0 : \beta = 0$ versus $H_a : \beta < 0$
(B) $H_0 : \beta = 0$ versus $H_a : \beta > 0$
(C) $H_0 : \beta > 0$ versus $H_a : \beta < 0$
(D) $H_0 : \beta \geq 0$ versus $H_a : \beta < 0$
(E) $H_0 : \beta < 0$ versus $H_a : \beta \geq 0$

399. A 95% confidence interval for the slope of a regression line is 1.04 ± 0.34. What is the value of the t-test statistic for $H_0 : \beta = 0$?

(A) 0.33
(B) 0.7
(C) 1.38
(D) 3.06
(E) The value cannot be determined from the given information

400. The t-test statistic for $H_0 : \beta = 0$ is 7.07. If the estimate for the slope is 1.51, find the standard error of the estimate S_b.

(A) 0.21
(B) 0.66
(C) 4.68
(D) 7.07
(E) The value cannot be determined from the given information

401. A scientist is studying the relationship between the percentage of a particular type of nutrient in the soil and the eventual height of a plant.

The regression equation is: Height = 5.31 + 3.34 Percentage

Predictor	Coef	St Dev	T	P
Constant	5.3083	0.3633	14.61	0.000
Percentage	3.345	1.316	2.54	0.019

$S = 0.831613$ R-Sq = 22.7%   R-Sq(adj) = 19.2%

Which of the following is true?
  I. In a test of $H_0 : \beta = 0$ versus $H_a : \beta \neq 0$, the null hypothesis would be rejected at the 5% level.
  II. The slope of the regression line is 3.345.
  III. The standard error of the slope is 0.3633.
  (A) I only
  (B) II only
  (C) III only
  (D) I and II only
  (E) I and III only

402. A marketing agency studied the relationship between the amount of money spent in a month on advertising (x) and the amount of sales in thousands of dollars (y). The calculated regression equation was $Y = 0.251X + 26251$. Given $H_0 : \beta = 0$, which of the following is correct at the 5% level?
  (A) Reject H_0 concluding that at the 5% level there is a linear relationship between the amount of money spent on advertising and sales.
  (B) Do not reject H_0 concluding that at the 5% there is a linear relationship between the amount of money spent on advertising and sales.
  (C) Reject H_0 concluding that at the 5% level there is no evidence of a linear relationship between the amount of money spent on advertising and sales.
  (D) Do not reject H_0 concluding that at the 5% level there is no evidence of a linear relationship between the amount of money spent on advertising and sales.
  (E) There is not enough information to conclude whether to reject H_0.

403. As part of a project analyzing the relationship between the square footage of homes and their final sale price in thousands of dollars, a student performed a regression analysis.

The regression equation is: Sale Price = 179 + 0.0289 SqFt

Predictor	Coef	St Dev	T	P
Constant	179.28	40.52	4.42	0.002
SqFt	0.02885	0.01888	1.53	0.165

S = 48.0769   R-Sq = 22.6%   R-Sq(adj) = 12.9%

Based on the computer output above, what is the value for the t-test statistic to test $H_0 : \beta = 0$?

(A) 1.53
(B) 4.42
(C) 3.06
(D) 8.84
(E) 40.52

404. The regression equation for two variables x and y is $Y = 1.293X + 41.029$ with $S_b = 2.967$. If the sample size used to determine this was 14, which of the following is the correct conclusion for the test $H_0 : \beta = 0$ at the 5% level?

(A) Reject H_0 concluding that at the 5% level there is no evidence of a linear relationship between x and y in the population.
(B) Reject H_0 concluding that at the 5% level there is evidence of a linear relationship between x and y in the population.
(C) Do not reject H_0 concluding that at the 5% level there is no evidence of a linear relationship between x and y in the population.
(D) Do not reject H_0 concluding that at the 5% level there is evidence of a linear relationship between x and y in the population.
(E) There is not enough information to conclude whether to reject H_0.

405. As part of a regression analysis on two variables, a student tests $H_0 : \beta = 0$ versus $H_a : \beta \neq 0$ and rejects the null hypothesis. Based on this result, which of the following is the correct conclusion?

(A) There is evidence of a linear relationship between the variables being studied.

(B) There is evidence of a positive linear correlation between the variables being studied.

(C) There is no evidence of a linear relationship between the variables being studied.

(D) There is evidence of a nonlinear relationship between the variables being studied.

(E) There is evidence that the regression line has a positive slope.

406. Suppose that as part of a linear regression analysis on two variables, a test of $H_0 : \beta = 0$ versus $H_a : \beta > 0$ does not reject the null hypothesis. Which of the following is the correct conclusion?

(A) There is evidence that the y-intercept of the regression line is positive.

(B) There is evidence that the y-intercept of the regression line is zero.

(C) There is evidence of a negative linear relationship between the variables.

(D) There is evidence of a positive linear relationship between the variables.

(E) There is no evidence of a positive linear relationship between the variables.

407. A 95% confidence interval for the slope of a regression line is (1.075, 1.992). Which of the following **best** interprets this interval?

(A) We are 95% confident that the true slope of the regression line is between 1.075 and 1.992.

(B) The probability is 0.95 that the true slope of the regression line is between 1.075 and 1.992.

(C) We are 95% confident that the true regression line will correctly predict y for x values between 1.075 and 1.992.

(D) For a given x value, the probability is 0.95 that the predicted y value will be between 1.075 and 1.992.

(E) We are 95% confident that predicted y values will be between 1.075 and 1.992 on average.

408. A scientist is studying the relationship between caffeine consumption (in mg) and scores on a test of alertness. She calculates a 90% confidence interval for the slope of the regression line to be (0.179, 0.234). Which of the following best interprets this interval?

(A) We are 90% confident that a 1 mg increase in caffeine consumption will result in an average increase of between 0.179 and 0.234 points on the test of alertness.

(B) The probability is 0.9 that participants in the study had between 0.179 and 0.234 mg of caffeine as part of the study.

(C) An increase in caffeine consumption of 1 mg will result in an increase of between 0.179 and 0.234 points on the test 90% of the time.

(D) There is a 90% chance that, on average, an increase of one point in the test of alertness will result in an increase in caffeine consumption of between 0.179 and 0.234 mg.

(E) The probability that the regression line will correctly predict the score on the test of alertness is between 0.179 and 0.234.

409. A regression analysis on two variables x and y resulted in a calculated 95% confidence interval for the slope of $(-1.27, 2.29)$. Based on this result, which of the following is not possible at the 5% level?

(A) There is a positive linear relationship between x and y.
(B) There is a negative linear relationship between x and y.
(C) There is no linear relationship between x and y.
(D) The null hypothesis $H_0 : \beta = 0$ will be rejected.
(E) The true slope of the regression line is 2.11.

410. A 95% confidence interval for the slope of a regression line is 2.715 ± 0.981. If this confidence interval is based on a sample of 25, find S_b.

(A) 0.476
(B) 0.474
(C) 0.981
(D) 2.029
(E) 3.696

411. Using a sample of 21 points, the calculated regression line is $\hat{y} = -1.215X + 15.211$. If $S_b = 0.8496$, find a 90% confidence interval for the true slope of the regression line.

(A) -1.215 ± 0.8496
(B) 15.211 ± 0.8496
(C) -1.215 ± 1.469
(D) 15.211 ± 1.729
(E) The confidence interval cannot be determined using the given information

412. A test of $H_0 : \beta = 0$ versus $H_a : \beta \neq 0$ does not reject the null hypothesis at the 5% level. If calculated from the same sample, which of the following is a possible 95% confidence interval for the true slope of the regression line?

(A) 2.651 ± 0.365

(B) 0.255 ± 0.365

(C) -1.712 ± 0.365

(D) 0.569 ± 0.365

(E) It is not possible to determine any properties of the 95% confidence interval based on the hypothesis test

413. A test of $H_0 : \beta = 0$ versus $H_a : \beta > 0$ rejects the null hypothesis at the 10% level. If calculated from the same sample, which of the following is a possible 90% confidence interval for the true slope of the regression line?

(A) 1.789 ± 0.514

(B) 0.355 ± 0.514

(C) -3.658 ± 0.514

(D) -0.822 ± 0.514

(E) It is not possible to determine any properties of the 5% confidence interval based on the hypothesis test

414. A test of $H_0 : \beta = 0$ versus $H_a : \beta < 0$ rejects the null hypothesis at the 1% level. If calculated from the same sample, which of the following is a possible 99% confidence interval for the true slope of the regression line?

(A) -2.833 ± 0.429

(B) 0.199 ± 0.429

(C) -1.658 ± 0.429

(D) 0.719 ± 0.429

(E) It is not possible to determine any properties of the 99% confidence interval based on the hypothesis test

415. As part of analyzing the relationship between two variables, it is determined that a 95% confidence interval for the true slope of the regression line is 1.7221 ± 0.7643. Using the same sample data, what would the result of a test of $H_0 : \beta = 0$ versus $H_a : \beta \neq 0$ be at the 5% level?

(A) Reject H_0 concluding that at the 5% level there is no evidence of a linear relationship between the variables.

(B) Reject H_0 concluding that at the 5% level there is evidence of a linear relationship between the variables.

(C) Do not reject H_0 concluding that at the 5% level there is no evidence of a linear relationship between the variables.

(D) Do not reject H_0 concluding that at the 5% level there is evidence of a linear relationship between the variables.

(E) There is not enough information to conclude whether to reject H_0.

416. As part of analyzing the relationship between two variables, it is determined that a 95% confidence interval for the true slope of the regression line is 2.391 ± 0.655. Using the same sample data, what would the result of a test of $H_0 : \beta = 0$ versus $H_a : \beta < 0$ be at the 5% level?

(A) Reject H_0 concluding that at the 5% level there is no evidence of a negative linear relationship between the variables.

(B) Reject H_0 concluding that at the 5% level there is evidence of a negative linear relationship between the variables.

(C) Do not reject H_0 concluding that at the 5% level there is no evidence of a negative linear relationship between the variables.

(D) Do not reject H_0 concluding that at the 5% level there is evidence of a negative linear relationship between the variables.

(E) There is not enough information to conclude whether to reject H_0.

417. As part of analyzing the relationship between two variables, it is determined that a 95% confidence interval for the true slope of the regression line is 1.492 ± 0.781. Using the same sample data, what would the result of a test of $H_0 : \beta = 0$ versus $H_a : \beta > 0$ be at the 5% level?

(A) Reject H_0 concluding that at the 5% level there is no evidence of a positive linear relationship between the variables.

(B) Reject H_0 concluding that at the 5% level there is evidence of a positive linear relationship between the variables.

(C) Do not reject H_0 concluding that at the 5% level there is no evidence of a positive linear relationship between the variables.

(D) Do not reject H_0 concluding that at the 5% level there is evidence of a positive linear relationship between the variables.

(E) There is not enough information to conclude whether or not to reject H_0.

418. An investigation was conducted into the relationship between hours spent asleep and a self-assessment of mood (on a scale of 1–100 points). Based on a confidence interval, the researchers determined they were 95% confident that, on average, one more hour of sleep led to an increase in the self-assessed mood rating of between 2.456 and 3.985 points. Which of the following is the slope of the least-squares regression line used to calculate this interval?

(A) 0.765

(B) 1.529

(C) 2.456

(D) 3.221

(E) 3.985

419. The p-value for the test of $H_0 : \beta = 0$ versus $H_a : \beta \neq 0$ is 0.025. If using the same sample data, which of the following is a possible 99% confidence interval for the true slope of the regression line?

(A) 1.512 ± 0.642

(B) -2.822 ± 0.642

(C) 0.588 ± 0.642

(D) 0.982 ± 0.642

(E) The sample size is necessary to determine a possible confidence interval

420. The t-test statistic for the test of $H_0 : \beta = 0$ versus $H_a : \beta \neq 0$ is 1.512. If using the same sample data, which of the following is a possible 90% confidence interval for the true slope of the regression line?

(A) 2.355 ± 0.952

(B) -1.862 ± 0.952

(C) 0.608 ± 0.952

(D) -0.129 ± 0.952

(E) The sample size is necessary to determine a possible confidence interval

421. A test for significance of regression is performed and it is determined that the regression is significant at the 5% level. Which of the following is a possible slope for the sample regression line?

(A) -5.13

(B) -0.897

(C) 1.99

(D) 5.13

(E) Any of the values above could be a possible slope

422. Which of the following tests should be used to determine whether a calculated regression line is significant?

(A) A linear regression t-test using the slope

(B) A one-sample t-test on the average y value

(C) A t-test using the y-intercept

(D) A two-sample t-test on the average x and y values

(E) A chi-squared test for independence

423. In a test of the significance of a regression analysis, the calculated p-value is 0.033. Which of the following statements must be true?
  I. The slope of the regression line is nonzero at the 5% level.
  II. The y-intercept of the regression line is nonzero at the 5% level.
  III. The probability of the regression line will correctly predict y-values as 3.3%.

  (A) I only
  (B) II only
  (C) III only
  (D) I and II only
  (E) I and III only

424. Which of the following alternative hypotheses should be used to determine whether the slope of a regression line is positive?

  (A) $H_a : \beta \geq 0$
  (B) $H_a : \beta > 0$
  (C) $H_a : \beta \leq 0$
  (D) $H_a : \beta < 0$
  (E) $H_a : \beta = 0$

425. Suppose that the least-squares regression line for a dataset is $Y = 2.324 + 3.197X$. Since the estimate of the slope is positive, which of the following must be true of a test of $H_0 : \beta = 0$ versus $H_a : \beta \neq 0$ at the 5% level?

  (A) The null hypothesis will be rejected suggesting there is a positive linear relationship between the variables.
  (B) The null hypothesis will be rejected suggesting there is no linear relationship between the variables.
  (C) The null hypothesis will not be rejected suggesting there is a positive linear relationship between the variables.
  (D) The null hypothesis will not be rejected suggesting there is no linear relationship between the variables.
  (E) It is not possible to determine whether the null hypothesis will be rejected with the given information.

426. If it is known that the calculated slope for a regression line is -2.931, which of the following must be true?

  (A) The y-intercept of the regression line is also negative.
  (B) A 95% confidence interval for the true slope will contain all negative values.
  (C) For $H_0 : \beta = 0$ versus $H_a : \beta < 0$, the null hypothesis will be rejected.
  (D) A 99% confidence interval for the true slope will contain some or all negative values.
  (E) For $H_0 : \beta = 0$ versus $H_a : \beta \neq 0$, the null hypothesis will not be rejected.

427. In a test of $H_0 : \beta = 0$ versus $H_a : \beta \neq 0$, the null hypothesis is rejected at the 5% level. Which of the following statements regarding the 95% confidence interval for β cannot be true?

(A) The 95% confidence interval contains all positive values.
(B) The 95% confidence interval contains all negative values.
(C) The 95% confidence interval contains zero.
(D) The 95% confidence interval contains 3.089.
(E) Properties of the 95% confidence interval cannot be determined with the given information.

428. The regression line for a 15-element dataset is $Y = -2.592X + 0.2399$. If $S_b = 1.91$, what is the correct conclusion for a test of $H_0 : \beta = 0$ versus $H_a : \beta \neq 0$ at the 10% level?

(A) The null hypothesis will be rejected suggesting that there is a linear relationship between the variables.
(B) The null hypothesis will be rejected suggesting that there is no evidence of a linear relationship between the variables.
(C) The null hypothesis will not be rejected suggesting that there is a linear relationship between the variables.
(D) The null hypothesis will not be rejected suggesting that there is no evidence of a linear relationship between the variables.
(E) There is not enough information to determine whether the null hypothesis will be rejected.

429. A regression line is calculated using a sample of 20 data points. If the regression line is $Y = 5.915 + 0.0328X$ and the standard error of the estimate of the slope is 0.072, what is the correct conclusion for a test of $H_0 : \beta = 0$ versus $H_a : \beta > 0$ at the 5% level?

(A) The null hypothesis will be rejected suggesting that there is a positive linear relationship between the variables.
(B) The null hypothesis will be rejected suggesting that there is no evidence of a positive linear relationship between the variables.
(C) The null hypothesis will not be rejected suggesting that there is a positive linear relationship between the variables.
(D) The null hypothesis will not be rejected suggesting that there is no evidence of a positive linear relationship between the variables.
(E) The null hypothesis will be rejected suggesting that there is a negative linear relationship between the variables.

430. A regression analysis on a sample dataset with 18 points calculates the point estimate of the true slope of the regression line to be -3.91 with a standard error of 0.84. Which of the following conclusions is **best** for a test of $H_0 : \beta = 0$ versus $H_a : \beta < 0$ at the 5% level?

(A) The null hypothesis will not be rejected suggesting that there is no evidence of a negative linear relationship between the variables.

(B) The null hypothesis will not be rejected suggesting that there is evidence of a negative linear relationship between the variables.

(C) The null hypothesis will be rejected suggesting that there is no evidence of a negative linear relationship between the variables.

(D) The null hypothesis will be rejected suggesting that there is evidence of a negative linear relationship between the variables.

(E) There is not enough information to determine whether the null hypothesis will be rejected.

431. A student found that the t-test statistic for a test of $H_0 : \beta = 0$ was -1.775. If the student used a sample of 21 and the point estimate for the true slope was -1.475, which of the following is the 95% confidence interval for the true slope of the regression line?

(A) -1.475 ± 1.739

(B) -1.475 ± 1.775

(C) -1.475 ± 3.72

(D) -1.475 ± 5.48

(E) The confidence interval cannot be found with the given information

432. Which of the following is not necessary to find a 95% confidence interval for the true slope of a regression line?

  I. The standard error of the estimate of the slope

 II. The point estimate of the slope

III. The point estimate of the y-intercept

(A) I only

(B) II only

(C) III only

(D) I and II only

(E) I, II, and III

433. A test of $H_0 : \beta = 0$ versus $H_a : \beta \neq 0$ determines that there is evidence of a linear relationship between the variables x and y at the 10% level. Which of the following is a possible p-value for this test?

(A) 0.06

(B) 0.11

(C) 0.82

(D) 1.15

(E) The p-value could be any of these values

434. A test of $H_0 : \beta = 0$ versus $H_a : \beta \neq 0$ determines that there is evidence of a linear relationship between the variables x and y at the 10% level. Given the sample size is 22, which of the following is a possible t-test statistic for this test?

(A) 0.09
(B) 1.22
(C) −1.17
(D) 1.83
(E) Either 1.22 or −1.17 could be the t-test statistic for this test

435. The t-test statistic for a test of $H_0 : \beta = 0$ versus $H_a : \beta \neq 0$ is 2.643 and $S_b = 1.144$. Which of the following could be the least-squares regression line?

(A) $Y = 2.643X + 8.174$
(B) $Y = 3.024X + 1.170$
(C) $Y = 1.144X + 2.451$
(D) $Y = 2.310X + 4.001$
(E) $Y = 0.255X + 2.643$

436. A 95% confidence interval for the true slope of a regression line is (−3.378, −0.936). Which of the following is the point estimate of the slope?

(A) −0.936
(B) −2.157
(C) −2.442
(D) −3.378
(E) −4.314

Use the following to answer **Questions 437–439.**

A student is studying the relationship between the number of hours students work and their final exam scores in statistics courses. A regression analysis resulted in the computer output below.

The regression equation is: Score = 84.0 − 0.420 × Hours

Predictor	Coef	St Dev	T	P
Constant	83.982	7.427	11.31	0.000
Hours	−0.4195	0.2605	−1.61	0.146

$S = 10.1695$ R-Sq = 24.5%   R-Sq(adj) = 15.0%

437. What is the p-value that should be used to test $H_0 : \beta = 0$ versus $H_a : \beta > 0$?

(A) 0
(B) 0.146
(C) 0.292
(D) 0.708
(E) 0.854

438. What is the t-test statistic that should be used to test $H_0 : \beta = 0$ versus $H_a : \beta \neq 0$?

(A) −3.22
(B) −1.61
(C) −0.805
(D) 5.655
(E) 11.31

439. If the student used a sample of 10, which of the following is the 95% confidence interval for the true slope of the regression line based on this dataset?

(A) −0.4195 ± 10.1695
(B) −0.4195 ± 23.451
(C) −0.4195 ± 0.2605
(D) −0.4195 ± 0.601
(E) −0.4195 ± 11.957

440. A professor is studying the relationship between the number of practice problems completed by students in the week before an assessment and the amount of time spent on the assessment in minutes. The professor used data from 15 students and some of the computer output from the analysis is shown below:

Predictor	Coef	St Dev	T	P
Constant	59.603	4.261	13.99	0.000
Problems	−0.23335	0.08764	−2.66	0.020

$S = 9.91193$ R-Sq = 35.3%   R-Sq(adj) = 30.3%

(a) What is the least-squares regression line that describes the relationship between the number of practice problems completed and time taken on the assessment? Define any variables used in the equation.

(b) Perform a hypothesis test to determine whether there is a linear relationship between the number of practice problems completed and the time taken on the assessment.

441. A sample is used to determine a least-squares regression line to describe the relationship between two variables x and y. The calculated line is $Y = -23.47 + 1.85X$.

(a) Describe the type of relationship you would expect to see between x and y in the sample data.

(b) Is there enough information to calculate a confidence interval for the true slope of the regression line? If so, calculate a 95% confidence interval. Otherwise, explain what other information would be required to find the interval.

442. A regression analysis on a sample of 23 16-ounce cereal box prices (in dollars) and sugar content (in grams) produced the following computer output.

The regression equation is: Price $= 2.22 + 0.140$ Sugar

Predictor	Coef	St Dev	T	P
Constant	2.2180	0.7596	2.92	0.019
Sugar	0.14011	0.07797	1.80	0.110

$S = 0.531706$ R-Sq $= 28.8\%$ R-Sq(adj) $= 19.8\%$

(a) Calculate and interpret a 95% confidence interval for the true slope of the regression line.

(b) Based on your interval in part (a), what would be the result of a test for the significance of the regression at the 5% level? Explain.

443. A 95% confidence interval for the true slope of a regression line is calculated to be -2.154 ± 3.012. This calculation is based on a sample of 15 (x,y) data points.

(a) Describe what type of correlation is possible in the population dataset.

(b) Would a 99% confidence interval calculation possibly change your description in part (a)? Explain.

444. A scientist is studying the effect of a new treatment against a viral infection specifically through the relationship between the time after infection the drug is administered (in minutes) and the time until the virus is no longer present (also in minutes). As part of this analysis, the scientist rejects the null hypothesis in a test of $H_0 : \beta = 0$ versus $H_a : \beta \neq 0$ at the 10% level.

(a) Interpret the result of the hypothesis test in terms of the original situation.

(b) Explain what must be true of a 90% confidence interval if it is calculated using the same dataset.

445. A student calculates a linear regression line based on a sample of 40 data
values to be $Y = 1.253X + 18.555$.

    (a) Describe what you would expect to see in a scatter plot of the
scientist's sample data.

    (b) What can you say would be true of a 95% confidence interval for the
slope of the population regression line? Justify your answer.

Inference for Categorical Data

446. The chi-squared distribution can assume:
- (A) Only positive values
- (B) Only negative values
- (C) Positive and negative values or 0
- (D) Only 0
- (E) None of the above

447. For any chi-squared goodness of fit problem, the number of degree of freedom is found by:
- (A) $n - k - 1$
- (B) $n + 1$
- (C) $n - 1$
- (D) $n + k$
- (E) None of the above

448. The test whether two variables or characteristics are related results in:
- (A) An ANOVA table
- (B) A chi-squared table
- (C) A contingency table
- (D) A scatter diagram
- (E) None of the above

449. In the chi-squared test, the null hypothesis is rejected when:
- (A) The difference between the observed and the expected frequencies is significant
- (B) The difference between the observed and expected frequencies occurs by chance
- (C) The difference between the observed and the expected frequencies is small
- (D) The computed chi-squared is less than the critical value
- (E) None of the above

450. What is the critical chi-squared value at the 0.05 level of significance for a goodness of fit if there are four categories?

(A) 9.49
(B) 11.14
(C) 9.35
(D) 7.81
(E) None of the above

451. If you were to sample 14 scores from a standard normal distribution, square each score, and sum the squares, how many degrees of freedom does the chi-squared distribution have that corresponds to this sum?

(A) 13
(B) 11
(C) 12
(D) 14
(E) None of the above

452. What is the mean of a chi-squared distribution with eight degrees of freedom?

(A) 8
(B) 9
(C) 72
(D) 7
(E) None of the above

453. Which chi-squared distribution looks most like a normal distribution?

(A) A chi-squared distribution with 0 df
(B) A chi-squared with 1 df
(C) A chi-squared distribution with 5 df
(D) A chi-squared distribution with 10 df
(E) None of the above

454. While comparing their grades, students evaluated the distribution of their four teachers who teach the same course. Which test would they employ?

(A) Chi-square test of independence
(B) Chi-square test of homogeneity
(C) Chi-square goodness of fit test
(D) Normal distribution
(E) None of the above

455. A chi-square test for goodness of fit is done on a variable with 14 categories. What is the minimum chi-squared value necessary to reject the null hypothesis at the 0.05 level of significance?

(A) 23.68
(B) 22.36
(C) 27.49
(D) 26.12
(E) None of the above

456. A local store sells baseball cards in packets of 100. Three types of players are represented in each package—rookies, veterans, and All-Stars. The company claims that 30% of the cards are rookies, 60% are veterans, and 10% are All-Stars. Cards from each group are randomly assigned to the packets. Suppose you bought a package of cards and counted the players from each group. What method would you use to test the store's claim?

(A) Chi-squared goodness of fit test
(B) Chi-squared test of homogeneity
(C) Chi-squared test for independence
(D) One-sample t-test
(E) None of the above

457. A public opinion poll surveyed a simple random sample of voters. Respondents were classified by gender and by voting preference (Republican, Democratic, or independent). Results are shown below.

	Voting Preferences			
	Republican	**Democrat**	**Independent**	**Row Total**
Male	200	150	50	400
Female	250	300	50	600
Column total	450	450	100	1000

If you conduct a chi-squared test for independence, what is the expected frequency count of male independents?

(A) 40
(B) 50
(C) 60
(D) 180
(E) 270

458. What test would you use to determine whether a set of observed frequencies differ from their corresponding expected frequencies?

(A) The t-test for dependent samples
(B) The z-test
(C) The F-test
(D) The chi-squared test
(E) None of the above

459. When testing for the equality between two proportions, what is the correct alternative hypothesis?

(A) $H_a : p_1 - p_2 \neq 0$
(B) $H_a : p_1 - p_2 < p_2 - p_1$
(C) $H_a : p_1 - p_2 > p_2 - p_1$
(D) $H_a : p_1 - p_2 = 0$
(E) None of the above

460. Directional hypothesis are possible with chi-square:

(A) When there are at least two degrees of freedom
(B) When the precise values of the expected frequencies have been predicted
(C) Never
(D) When you have a 2×2 design
(E) None of the above

461. Which of the following values cannot occur in a chi-square distribution?

(A) 100
(B) 38.4
(C) 0.79
(D) 0.05
(E) −2.67

462. What other name is used for a contingency table?

(A) An ANOVA table
(B) A histogram
(C) A cross-classification table
(D) A z-test
(E) None of the above

463. A chi-square test involves a set of counts called "expected counts." What is the best description of the expected counts?

    (A) Hypothetical counts that would occur if the alternative hypothesis were true

    (B) Hypothetical counts that would occur if the null hypothesis were true

    (C) The actual counts that did occur in the observed data

    (D) The long-run counts that would be expected if the observed counts are representative

    (E) None of the above

464. The following data have been collected on the respondents' favorite type of sport:

	Individual Sport	Team Sport	Neither
Male	15	17	18
Female	4	32	17

If the expected frequencies rule for chi-square had been violated by the data, which categories could be combined together usefully in an attempt to increase the expected frequencies?

    (A) Individual sport, team sport, and neither

    (B) Males and females

    (C) Individual sport and team sport

    (D) All of the above

    (E) None of the above

465. A chi-squared test is appropriate to use for:

    (A) Scores

    (B) Ranks

    (C) Percentages

    (D) All of the above

    (E) None of the above

466. On average a standard nickel battery for cell phones lasts 200 minutes on a single charge. The standard deviation of the population's charge duration is 4 minutes. Suppose the quality control department selected a random sample of 20 batteries. The standard deviation of the selected sample is 12 minutes. What would be the chi-square statistic represented by this test?

    (A) 94

    (B) 188

    (C) 81

    (D) 48

    (E) None of the above

467. For the following two-way table, compute the value of χ^2.

	y	z
w	5	25
x	10	15

(A) 1.45
(B) 4.167
(C) 5.2
(D) 5.55
(E) None of the above

468. A toy company published an advertisement in the local newspaper of a major city and in a kids' magazine to see which medium generated more advertising viewership. A random sample of families was taken and each family was asked to indicate where they had seen the product's advertisement. The results are summarized below.

	Magazine	Newspaper
Excellent	42	48
Good	25	44
Fair	40	26
Never seen	52	43

The advertiser decided to use a chi-square test to see whether there was a relationship between the newspaper and the magazine scores. What would be the degrees of freedom for this test?

(A) 3
(B) 4
(C) 7
(D) 8
(E) None of the above

469. A physical education teacher at a middle school wanted to observe whether there was any association between the weight of a male student and how much weight he could lift. The following data were collected and a chi-square test was performed. The output indicated a test statistic of 1.745 and a p-value of 0.1865.

Total Weight Lifted (pounds)	Student Weight (pounds)		Total
	Under 125	Over 125	
Under 150	44	43	87
Over 150	25	38	63
Total	69	81	150

Based on the table above, which of the following is a correct statement?

(A) Since the value of the test statistic is so small, there is a significant relationship between the weight of an individual and the weight he can lift.

(B) The correlation coefficient from a linear regression analysis for these data would be less than 0.20.

(C) It is not possible to run a chi-square test on these data since the conditions for conducting this test are not met.

(D) The p-value of this test is large, which indicates that no significant relationship exists between the weight of an individual and the weight he can lift.

(E) The p-value of this test is small, which indicates that a significant relationship exists between the weight of an individual and the weight he can lift.

470. When performing a hypothesis test for the slope of a linear regression, which is the appropriate test to use?

(A) One-sample z-test

(B) One-sample t-test

(C) Two-sample z-test

(D) Two-sample t-test

(E) Chi-square test for goodness of fit

471. The following computer output shows results for a least-squares regression. Which of the following would be the appropriate null and alternative hypothesis for a test of significance about the slope of the regression line?

Predictor	Coef	SE Coef	T	P
Constant	−257.19	14.29	−18	0.0000
Chest.G	12.5749	0.3838	32.76	0.0000

(A) $H_0 : \beta = 0$ $H_a : \beta \neq 0$
(B) $H_0 : \beta = 0.3838$ $H_a : \beta \neq 0.3838$
(C) $H_0 : \beta = 12.5749$ $H_0 : \beta \neq 12.5749$
(D) $H_0 : \beta = 32.76$ $H_0 : \beta \neq 32.76$
(E) $H_0 : \beta = 0.3838$ $H_0 : \beta > 0.3838$

472. A highway superintendent states that five bridges into a city carry vehicles in the ratio 2:3:3:4:6 during the morning rush hour. A highway study of a simple random sample of 600 cars indicates that 720, 970, 1,013, 1,380, and 1,917 cars use the five bridges, respectively. Can the superintendent's claim be rejected at the 2.5% or 5% level of significance?

(A) There is sufficient evidence to reject the claim at either of these two levels.
(B) There is sufficient evidence to reject the claim at the 5% level but not at the 2.5% level.
(C) There is sufficient evidence to reject at the 2.5% level but not at the 5% level.
(D) There is not sufficient evidence to reject the claim at either of these levels.
(E) There is not sufficient information to answer this question.

473. An economist is studying the distribution of income. He wants to see whether the income distribution in the states of New Jersey, Washington, Texas, and Georgia are comparable. Which test would he employ?

(A) t-test
(B) z-test
(C) χ^2 test for homogeneity
(D) p-test
(E) None of the above

474. A two-way table of counts is analyzed to examine the hypothesis that the row and column classifications are independent. There are three rows and four columns. The degrees of freedom value for the chi-square statistic is:

(A) 12
(B) 11
(C) 6
(D) The minimum of $n_1 - 1$ and $n_2 - 1$
(E) None of the above

475. In a recent study of accident records at a large engineering company in Europe reported the following number of injuries in each shift for one year.

Shift	Morning	Afternoon	Night
Number of injuries	879	1067	1200

Is there sufficient evidence to say that the number of accidents during the three shifts are not the same? Test at the 0.05, 0.01, and 0.001 levels.

(A) There is sufficient evidence at all three levels to say that the number of accidents during each shift are not the same.
(B) There is sufficient evidence at both 0.05 and 0.01 levels but not at the 0.001 level.
(C) There is sufficient evidence at the 0.05 level but not at the 0.01 and 0.001 levels.
(D) There is sufficient evidence at the 0.001 level but not at the 0.05 and 0.01 levels.
(E) None of the above

Questions 476–478 refer to the following passage and chart:

A single sample from the population of hikers was taken and the hikers were sorted into the cells by level of experience (novice, experienced) and preferred hiking direction if lost (downhill, uphill, or remain where they were). The null hypothesis is that there is no association between the two variables. As in all hypothesis tests, we use the null hypothesis to state the expected values of the proportions. The table below should be used to answer the next three questions.

	Uphill	Downhill	Remain
Novice	20	50	50
Experienced	10	30	40

476. If a hiker from this sample were selected at random, what is the probability that he or she would be a novice?

(A) $\dfrac{100}{240}$

(B) $\dfrac{120}{200}$

(C) $\dfrac{100}{180}$

(D) $\dfrac{100}{80}$

(E) None of the above

477. If a hiker from this sample were selected at random, what is the probability that he or she would go uphill?

(A) $\dfrac{30}{200}$

(B) $\dfrac{80}{200}$

(C) $\dfrac{90}{200}$

(D) $\dfrac{30}{100}$

(E) None of the above

478. Based on these two probabilities, if the level of hiking experience and the direction a hiker would travel if lost are independent, what is the probability that a hiker selected at random would be a novice who would head uphill?

(A) $\dfrac{120}{200}$

(B) $\dfrac{30}{200}$

(C) $\dfrac{9}{100}$

(D) $\dfrac{9}{200}$

(E) None of the above

479. A building inspector inspects four major construction sites every day. In a sample of 100 days, the number of times each site passed inspection is shown in this chart.

Site	A	B	C	D
Passed	45	33	67	83

Test the hypothesis that Sites C and D pass inspection twice as often as Sites A and B; that is, that the frequency of passing inspection has the ratio 1:1:2:2.

(A) There is sufficient evidence at both the 10% and 5% significance levels that there is not a good fit with the indicated ratio.

(B) There is sufficient evidence at the 10% level, but not at the 5% level, that there is not a good fit with the indicated ratio.

(C) There is sufficient evidence at the 5% level, but not at the 10% level, that there is not a good fit with the indicated ratio.

(D) There is not sufficient evidence at either the 10% or 5% significance levels that there is not a good fit with the indicated ratio.

(E) There is not sufficient information to answer the question.

480. Students are comparing the grade distributions of three professors who teach the same course. Which test would they employ?

(A) χ^2 test for independence
(B) χ^2 test for goodness for fit
(C) χ^2 test for homogeneity
(D) t-test
(E) None of the above

481. The random variable χ^2 has values in what range?

(A) All real numbers
(B) -1 to 1
(C) 0 to 1
(D) Nonnegative values only
(E) It depends on the degrees of freedom

482. When performing a hypothesis test for the slope of a linear regression, which is the appropriate test to use?

(A) One-sample z-test
(B) One-sample t-test
(C) Two-sample z-test
(D) Two-sample t-test
(E) Chi-squared test for goodness of fit

483. It is generally agreed that the use of the chi-squared distribution is appropriate when:

(A) The sample size is at least 30
(B) The sample size is large enough so that all of the observed cell counts is at least five
(C) The sample size is large enough so that all of the expected cell counts is at least five
(D) The sample size is large enough so that at least one of the expected cell counts is at least five
(E) The sample size is large enough so that the average of the expected cell counts is at least five

484. A two-way table of counts is analyzed to examine the hypothesis that the row and column classifications are independent. There are three rows and four columns. The degrees of freedom value for the chi-square statistic is:

(A) 12
(B) 11
(C) 6
(D) The minimum of $n_1 - 1$ and $n_2 - 1$
(E) None of the above

485. The χ^2 distribution is:

(A) Symmetric
(B) Skewed left
(C) Bimodal
(D) Unimodal
(E) Skewed right

486. A specific chi-square distribution is specified by one parameter, called the:

(A) Normal curve
(B) Degrees of freedom
(C) t-test
(D) z-test
(E) None of the above

487. A χ^2 goodness of fit test is performed on a sample of 400 16-year-old students to see if there whether a relationship between the number of them getting their licenses and their birth months. The χ^2 value is determined at 23.5. The p-value for this test is:

(A) $0.001 < p < 0.005$
(B) $0.02 < p < 0.025$
(C) $0.025 < p < 0.05$
(D) $0.01 < p < 0.02$
(E) $0.05 < p < 0.10$

488. Using the goodness of fit test, a random sample of the top 10 laundry detergents are chosen to show individual preferences. If a sample size of 100 is chosen, which of the following would prove the point at the 0.01 level of significance?

(A) $\chi^2 > 135.8$
(B) $\chi^2 > 23.21$
(C) $\chi^2 > 135.8$
(D) $\chi^2 > 21.67$
(E) $\chi^2 > 140.2$

489. During fall birds migrate from north to south, and a survey shows that for four particular species the ratio is 1:3:3:9. Suppose a sample of 400 birds contained 220, 780, 730, and 2,230 birds of each species, respectively. Is there sufficient evidence to reject what the survey claims?

(A) The test proves what the survey claims.
(B) The test proves that the survey is false.
(C) The test does not give sufficient evidence to reject the claim.
(D) The test gives sufficient evidence to reject the claim.
(E) The test is inconclusive.

490. ABC Toys sells baseball cards in packages of 100. Three types of players are represented in each package—rookies, veterans, and All-Stars. The company claims that 30% of the cards are rookies, 60% are veterans, and 10% are All-Stars. Cards from each group are randomly assigned to packages.

Suppose you bought a package of cards and counted the players from each group. What method would you use to test ABC's claim that 30% of the cards are rookies, 60% are veterans, and 10% are All-Stars?

(A) Chi-square goodness of fit test
(B) Chi-square test for homogeneity
(C) Chi-square test for independence
(D) One-sample t-test
(E) Matched-pairs t-test

491. Which of the following tests should be used to determine whether two factors are related to each other?

(A) A chi-squared goodness of fit test
(B) A two-proportion z-test
(C) A two-sample t-test
(D) A chi-squared test of independence
(E) A two-sample F-test

492. Using the given contingency table, determine whether there is evidence that the type of vote is dependent on gender at the 5% level.

	Voted Yes	Voted No
Male	32	28
Female	25	40

(A) The null hypothesis is rejected, which implies there is evidence that the type of vote is dependent on gender.
(B) The null hypothesis is not rejected, which implies there is evidence that the type of vote is dependent on gender.
(C) The null hypothesis is rejected, which implies that there is not evidence that the type of vote is dependent on gender.
(D) The null hypothesis is not rejected, which implies that there is not evidence that the type of vote is dependent on gender.
(E) There is not enough information to determine whether the null hypothesis would be rejected.

493. Which of the following must be true to perform a chi-squared test of independence?
  I. The observations are based on a random sample.
 II. The observed frequencies are all greater than 30.
III. The data are presented in a table format.

(A) I only
(B) II only
(C) III only
(D) I and II only
(E) I and III only

494. Given the contingency table below, determine the expected frequency of cell 1, 3.

	A′	B′	C′
A	110	120	115
B	100	98	101
C	85	82	88

(A) 83.7
(B) 85
(C) 115
(D) 116.7
(E) 899

495. Employees at a large company are asked to rate their experiences with the company so far from 1 to 5. For part-time employees included in the sample, 10 rated their experiences as a 1, 12 a 2, 15 a 3, 27 a 4, and 11 a 5. Among full-time employees, 27 rated their experiences as a 1, 35 a 2, 45 a 3, 12 a 4, and 30 a 5.

(a) Create a contingency table to represent these sample data.

(b) Is there evidence that the rating is dependent on whether the employee is employed full time or part time? Justify your answer.

496. In a recent survey taken about the major channels on TV, it was found that channels 2, 3, 4, and 5 captured 30%, 25%, 20%, and 25% of the audience, respectively. During the first week of the new season, 500 viewers were interviewed.

(a) If viewer preferences have not changed, what number of people are expected to watch each channel?

(b) Suppose that the actual observed numbers are as follows:

	Channel			
	2	**3**	**4**	**5**
Observed Number	139	138	112	111

Do these numbers indicate a change? Are the differences significant?

497. Computer software generated 500 random numbers that should look like they are from the uniform distribution on the interval 0 to 1. They are categorized into five groups:

(1) Less than or equal to 0.2

(2) Greater than 0.2 and less than or equal to 0.4

(3) Greater than 0.4 and less than or equal to 0.6

(4) Greater than 0.6 and less than or equal to 0.8

(5) Greater than 0.8

The counts in the five groups are 103, 90, 110, 89, and 95, respectively.

(a) The probabilities for these five intervals are all the same. What is this probability?

(b) Compute the expected count for each interval for a sample of 500.

(c) Perform the goodness of fit test and summarize your results. Show the entire calculation of the χ^2 statistic.

498. A social scientist sampled 140 people and classified them according to income level and whether they played a state lottery in the last month. The sample information is reported below. Is it reasonable to conclude that playing the lottery is related to income level? Use a 0.05 significance level.

	Income			
	Low	Middle	High	Total
Played	46	28	21	95
Did not play	14	12	19	45
Total	60	40	40	140

(a) What is this table called?
(b) State the null and the alternative hypothesis.
(c) Determine the chi-squared value.

499. A public opinion poll surveyed a simple random sample of 1,000 voters. Respondents were classified by gender) and by voting preference (Republican, Democratic, or independent). Results are shown in the contingency table below.

	Voting Preferences			
	Republican	Democrat	Independent	Total
Male	200	150	50	400
Female	250	300	50	600
Total	450	450	100	1,000

(a) State the null and alternative hypothesis that test for a gender gap in voting preference.
(b) Do the men's voting preferences differ significantly from the women's preferences? Use a 0.05 level of significance.

500. A media company plans to publish a special edition of a newspaper. Past experience shows that the number of newspapers the company will sell is described by the following table:

Newspaper	210k–220k	220k–230k	230k–240k	240k–250k	250k–260k	260k–270k	270k–280k	280k–290k
Probability	0.01	0.07	0.14	0.25	0.24	0.15	0.10	0.04

(a) What is the probability that the company will sell less than 250,000 newspapers?

(b) The cost of selling this special edition is $100,000 for up to 260,000 newspapers and an additional flat cost of $10,000 if more are sold. What is the expected sum the company will spend to sell the special edition?

(c) Given that the company spends the flat extra cost of $10,000, what is the probability that the company sells between $270,000 and $280,000 newspapers?

ANSWERS

Chapter 1: Overview of Basic Statistics

1. (B) AKC dog breed is a categorical designation used to specify group characteristics and, unlike quantitative data, cannot be used in meaningful computations.

2. (A) Radiation levels are numerical measurements that can be used in meaningful computations.

3. (D) This is a count of individual items and takes on only positive whole number values, which makes it discrete.

4. (C) The value of the NYSE is a measurement that can take on any value in an interval, unlike discrete data, which can only take on whole number values.

5. (A) The average, which is used to describe the center of data, is an example of a descriptive statistic that summarizes data.

6. (B) This study of pigeons incorporates hypothesis testing to determine whether there is a significant difference in the blood parasite concentrations in urban pigeons grouped by color.

7. (D) This measure was calculated from a sample taken from a population.

8. (D) This measure was calculated from a population that is specifically defined; no sampling was done. A parameter is a characteristic of an entire population.

9. (A) The farmer would experiment by planting the three varieties in separate plots under like conditions, harvest the crops, and compare the yields. The other choices describe situations involving data collection using a census or survey.

10. (A) Gallup would randomly sample the population using a survey designed to identify individuals who see themselves as "thriving." The other choices describe experiments or simple calculations of a population parameter.

11. (B) This is generated from a random process (flipping a coin, rolling a die, spinning a spinner, etc.). The random variable takes on the values of the outcome from a process that cannot be predicted.

12. (C) To avoid bias, we collect data carefully. If we do not, then our data will contain too much error that cannot be attributed to randomness. Random error is considered outside our control.

13. (B) To measure the central tendency of data, use the mean, median, or mode.

14. (E) To measure the spread of data, use the variance and standard deviation.

15. (B) When the population is so large that we cannot compute parameters, collecting a representative sample of the population and calculating the sample statistics allows us to estimate the parameters, i.e., we make an inference from the sample to the population.

16. (C) Random sampling is the common technique because it uses a random process to select the members of the sample, avoiding human bias.

17. (C) Scatter plot. The scatter plot is a two-dimensional graph that allows you to plot bivariate data points and view the trend pattern of the point cloud.

18. (B) The histogram provides a view of the shape of univariate data.

19. (D) The treatment was receiving eyeglasses. This treatment is compared with the results from the control group who did not receive eyeglasses.

20. (A) This is an observational study. The researcher collected data by making observations from the videotapes. The other choices describe studies that use a control versus treatment group experimental design or conduct surveys to make generalizations about the population.

Chapter 2: One-Variable Data Analysis

21. (E) Uniform describes the shape of this histogram because the bars are approximately the same height, which represents approximately equal frequencies.

22. (B) Bimodal describes the shape of this histogram because there are exactly two clearly defined peaks of the same height, which represent two classes with the greatest frequencies.

23. (B) Normal describes the shape of this histogram because it is unimodal and approximately symmetrical with thinning tails in the extreme left and right.

24. (C) The mode is 14 since it has the highest frequency, and the median is 13.5 = (13 + 14)/2 since the number of data points is even.

25. (D) The modal group is two, which represents 20–29 and has the most data values in it; the median is 34 = (34 + 34)/2 since the number of data points is even.

26. (E) The boundaries are midway between the bar labels: left boundary: (7.5 + 10)/2 = 8.75; right boundary: (10 + 12.5)/2 = 11.25.

27. (B) This distribution is right skewed. When a distribution is right skewed, the mean is greater than the median and the median is greater than the mode.

28. (C) This distribution is left skewed. When a distribution is left skewed, the mean is less than the median and the median is less than the mode.

29. (C) Any normal distribution that is standardized becomes the standard normal distribution with a mean of 0 and standard deviation of 1.

30. (D) The median is the most resistant statistic because it stays in the center and is not influenced by outliers.

31. (B) Mean is $(3 + 8 + 10 + 3 + 12 + 7 + 10)/7 = 7.571$. Variance is $[(3 - 7.571)^2 + (8 - 7.571)^2 + (10 - 7.571)^2 + (3 - 7.571)^2 + (12 - 7.571)^2 + (7 - 7.571)^2 + (10 - 7.571)^2]/(7 - 1) = 12.286$. Standard deviation is square root $(12.286) = 3.5$.

32. (C) Standard deviation is a measure of spread so it is sensitive to the spread of data and is influenced by extreme values. It is independent of the mean because the spread does not depend on where the mean is and the mean itself is dependent on the data.

33. (D) Reading from the graph, Min: 3; Q1: 5.75; Med: 9.5; Q3: 13.75; Max: 16. Calculating the interquartile range $= Q3 - Q1 = 13.75 - 5.75 = 8$.

34. (E) Interquartile range (IQR) is $10.5 - 4.5 = 6$. To find the boundary for mild outliers, multiply the IQR by $1.5:6(1.5) = 9$. Any values that are less than $Q1 - 9$ or greater than $Q3 + 9$ are considered mild outliers (but not more than 3(IQR), which is considered extreme). 20 is a mild outlier because it is greater than $10.5 + 9 = 19.5$ and less than $10.5 + 18 = 28.5$.

35. (C) Outliers are natural and rare but should be investigated because they may indicate that an error occurred in the process of experimentation or recording data.

36. (C) z-score $= (84 - 72)/4.5 = 2.67$. This means that her grade is 2.67 standard deviations above the mean. She is in the top 1% of the class since the area under the standard normal curve to the left of $z = 2.67$ is $>99\%$.

37. (C) Whitaker: $z = (45 - 42.5)/7.6 = 0.33$; Mirren: $z = (61 - 35)/9.7 = 2.68$. Whitaker's age was about average within one standard deviation of the mean while Mirren's was well above average close to three standard deviations above the mean.

38. (B) The empirical rule states that 68% of the data are plus or minus one standard deviation from the mean.

39. (E) According to Chebyshev's rule, $(1 - 1/k^2)\%$ are within k standard deviations from the mean for any distribution. $(1 - 1/2^2) = 0.75$ or 75%. $0.75(5,250) = 3,937.5$. To be sure, round down to 3,937.

40. (C) The empirical rule applies only to normally distributed data while Chebyshev's rule applies to any distribution and is a rough estimate.

41. (B) All probability density curves are mathematical models of probability distributions. Probability distributions are proportional distributions of data and the area underneath them sums to 1. They are also always above the horizontal axis because probability is nonnegative.

42. (B) I and III only. By definition, a probability density curve is a *model* and a mathematical function. The constraints are that the area between the curve and the x-axis is equal to 1 and that the function is nonnegative. It is not necessary for the function itself to be less than 1. An example is the uniform distribution where the probability density curve is constant and on the interval [0, 1] has probability density $f(x) = 1$. I and III are above the horizontal axis and the areas sum to 1. This can be verified by using the area formula for a triangle: $A = \frac{1}{2}bh$. For I, $= \frac{1}{2}bh = \frac{1}{2}2(1) = 1$. For III, the area is the sum of a rectangle and a triangle.

$$A = A_1 + A_2 = 2(0.25) + \frac{1}{2}1(1) = 1.$$

43. (E) III only. By definition, a probability density curve is a *model* and a mathematical function. The constraints are that the area between the curve and the x-axis is equal to 1 and that the function is nonnegative. It is not necessary for the function itself to be less than 1. An example is the uniform distribution where the probability density curve is constant and on the interval [0, 1] has probability density $f(x) = 1$. In III, the triangular area is symmetrical with respect to 1.

44. (E). The five-number summary for Bonds: Min: 16; Q1: 25; Med: 35; Q3: 41.5; Max: 73. The five-number summary for Aaron: Min: 13; Q1: 28; Med: 38; Q3: 44; Max: 47. These summaries are correctly represented in the box plot in choice **(E)**.

45. (E) The dataset contains 31 data points. Minimum is 13; first quartile is the 8th data point or 23; median is the 16th data point or 27; third quartile is the 24th data point or 34; and maximum is 51.

46. (A) There are more cat owners than dog owners. There are more cats than dogs. This can be estimated by the relative heights of the bars labeled 1, 2, 3, 4, and 5. Notice that there are more households that own greater quantities of cats.

47. (C) When all data values are increased by a constant, the mean, $\bar{X} = \frac{1}{n}\sum_{i=1}^{n}X_i$, increases by that constant as shown here: $\bar{X} = \frac{1}{n}\sum_{i=1}^{n}(X_i + C) = \frac{1}{n}\sum_{i=1}^{n}X_i + \frac{1}{n}\sum_{i=1}^{n}C = \frac{1}{n}\sum_{i=1}^{n}X_i + C$. The standard deviation does not change when all data values are increased by the same amount because the average distance from the mean is unchanged.

48. (B) The best *estimate* is 2. The true value is $\bar{X} = \frac{5(0) + 16(1) + 29(2) + 11(3)}{5 + 16 + 29 + 11} = 1.8$.

49. (C) From the pie chart, the frequency distribution of the data can be reconstructed. A total of 2 made one trip, 9 made two trips, 15 made three trips, 7 made four trips, and 3 made five trips. $\bar{X} = \frac{2(1) + 9(2) + 15(3) + 7(4) + 3(5)}{2 + 9 + 15 + 7 + 3} = 3.0$. To calculate s, the sum of squares: $SS_X = 2(1-3)^2 + 9(2-3)^2 + 15(3-3)^2 + 7(4-3)^2 + 3(5-3)^2 = 36$. $s = \sqrt{\frac{SS_X}{n-1}} = \sqrt{\frac{36}{35}} = 1.0$. Statistics are rounded to tenths since the precision of the data was in whole numbers.

50. (B) When one data value is increased and another is decreased by a constant, the mean, $\bar{X} = \frac{1}{n}\sum_{i=1}^{n} X_i$, remains the same: $\bar{X} = \frac{1}{n}((X_1 - C) + X_2 + \cdots + (X_{MED} + C) + \cdots + X_n) = \frac{1}{n}\sum_{i=1}^{n} X_i$. The standard deviation increases since it is sensitive to the range, which has increased by c. $SS_x = \sum X^2 - \frac{(\sum X)^2}{n}$. Note that $\sum X$ will not change because the net change from adding and subtracting 12 is 0; however, $\sum X^2$ does increase because $(X - C)^2 = X^2 - 2CX + C^2$ and $(X + C)^2 = X^2 + 2CX + C^2$ and the new SS_x is greater by $2C(X_{MED} - X_{MIN}) + 2C^2$.

51. (D) Minimum is 2.9; maximum is 4.2; median = (3.5 + 3.6)/2 = 3.55; first quartile = 3.15; and third quartile = 3.9.

52. (C) Reading from the vertical axis the percentages and down to the horizontal axis, $P_{20} = 96$; $P_{50} = 102$; $P_{80} = 106$.

53. (C) The z-score associated with 20% to the left of it is −0.84. To find the height associated with that z-score: $z\sigma + \mu = -0.84(2.6) + 69.6 = 67.4$. That means that 20% of the population is shorter than 68 inches tall.

54. (D) $z = \frac{X - \mu}{\sigma}$. Calculating for each height: 2.08, −2.54, and 3.62, respectively, shows that all are greater than two standard deviations from the mean.

55. (B) The area under the normal distribution associated with 67 inches is 0.1587 or the 16th percentile. Sixteen percent of males between the ages 20–29 are shorter than 67 inches.

56. (A) Mean: (9 + 7 + 8 + 6 + 9 + 12 + 11 + 5 + 9 + 10)/10 = 8.6; Median: (9 + 9)/2 = 9; Mode: 9, the most frequent data point occurring three times.

57. (B) The $SS_X = 42.4$. Variance, $\sigma^2 = \frac{42.4}{9} = 4.71$. Standard deviation, $\sigma = \sqrt{4.71} = 2.2$. Range = Max − Min = 12 − 5 = 7.

58. (B) $\bar{X} = \frac{1}{n}\sum_{i-1}^{n} X_i = \frac{1}{50}\sum Xf = \frac{0(12) + 1(8) + 2(17) + 3(7) + 4(3) + 5(1) + 6(2)}{50} = 1.84$.

59. (A) $s = \sqrt{\frac{SS_X}{n-1}}$ where $SS_x = \sum X^2 - \frac{(\sum X)^2}{n} = 114.72$. Therefore $s = \sqrt{\frac{114.72}{49}} = 1.53$.

60. (A) The area under the normal curve associated with mean $106.50 and standard deviation $3.85 per month and between $100 and $115 is 0.9409. The z-scores for 100 and 115 are −1.69 and 2.21, respectively. The area under the standard normal curve between those scores is 0.9409.

61. **(B)** The interquartile ranges for the datasets are 6 and 2, respectively. There are no outliers in Dataset 1 so the mean of 6.2 would be the best measure of central tendency. 15 is an extreme outlier in Dataset 2, so the median 5.5 would be the best measure of central tendency since the mean is more sensitive to outliers.

62. **(A)** The z-score with area of 0.9 to the left of it is 1.28 which means that the 90th percentile or above is 1.28 standard deviations above the mean.

63. **(A)** The z-score with area of 0.7 to the left of it is 0.52 which means that the 70th percentile or above is 0.52 standard deviations above the mean.

64. **(D)** The z-score associated with a test score of 65 is $\dfrac{65-72}{4.2} = -1.67$. The area to the left of a z-score of -1.67 is 0.0475 or approximately the 5th percentile.

65. **(D)** The data point is at the point of inflection on the bell curve which is one standard deviation above the mean or $X = 67 + 5.1 = 72.1$.

66. **(a)** If there are outliers, particularly extreme outliers, the median is less sensitive and more resistant than the mean. That may explain why median salaries would be reported rather than average salaries.

   (b) From the summary data, extracting the specialists and finding the median shows that, for men, it is $\dfrac{148+171}{2} = 159.5$. For women, it is $\dfrac{143+126}{2} = 134.5$.

	Men Med	Wom Med	Diff
Podiatrists	89	74	14
Pediatricians	146	112	34
Obstetricians and Gynecologists	148	143	5
Psychiatrists	171	126	45
Surgeons	200	164	36
Anesthesiologists	247	194	53

This is lower than what was reported; additionally, the wage gap for these two medians is about 16 cents, which is less than the reported 23 cents for 2009. $\dfrac{159.5-134.5}{159.5} = \dfrac{x}{100}, x = 16.$

67. (a) Stem-and-leaf of No Incentive N = 10
Leaf Unit = 1.0

1	7	3
3	8	59
(3)	9	346
4	10	38
2	11	5
1	12	3

Stem-and-leaf of Incentive N = 10
Leaf Unit = 1.0

4	8	0259
5	9	0
5	10	4
4	11	6
3	12	
3	13	125

(b) The box plots below indicate that neither group has any outliers. There are no data points in either group that are more than $1.5(IQR)$ above Q3 or more than $1.5(IQR)$ below Q1.

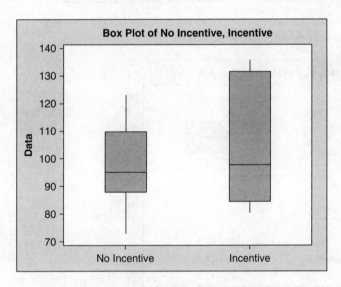

(c) The mean would be an appropriate measure for both groups in the absence of outliers. The median would also be an appropriate measure of central tendency.
No Incentive: mean = 97.9, median = 95.0
Incentive: mean = 105.0, median = 98.0

(d) Quartile Three is the 75th percentile and here it is reported as 109.75 for the No Incentive group and 131.39 for the Incentive Group. From the stem-and-leaf-plot, 108 and 131, respectively.

68. (a) Histograms for each individual variable are shown below.

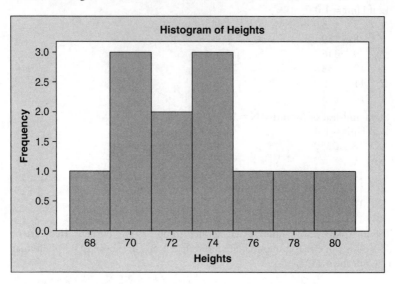

(b) The box plots indicate that there are no outliers.

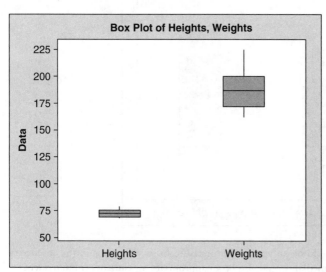

(c) The means would be an appropriate measure of central tendency in the absence of outliers. Mean height = 72.75 inches. Mean weight = 187.83 pounds.

(d) The height for the 25th percentile is the same as the first quartile. That is 69.5, the median from the lower half of the data.

69. (a)

Classes	Frequency
144.5–165	8
165.5–186	7
186.5–207	3
207.5–228	1
228.5–249	1
	20

(b) A stem-and-leaf plot or a histogram would be an appropriate graphical display.
Stem-and-leaf of Water consumed $n = 20$

Leaf Unit = 1.0

```
  4  | 14   5669
  5  | 15   1
  9  | 16   0237
 (5) | 17   34578
  6  | 18   08
  4  | 19   25
  2  | 20
  2  | 21
  2  | 22   4
  1  | 23
  1  | 24   4
```

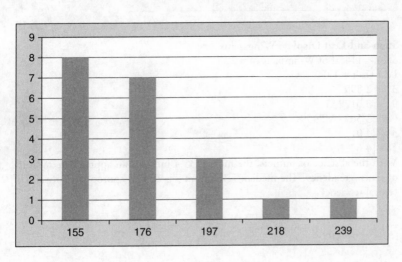

(c) Due to the outlier, the median—173.5 gallons per day—would be an appropriate measure of central tendency.

Descriptive Statistics: Water consumed

Variable	N	N*	Mean	SE Mean	St Dev	Minimum	Q1	Median	Q3
Water consumed	200		174.45	5.73	25.64	145.00	153.25	173.50	186.00

Variable Maximum
Water consumed 244.00

(d) The box plot below indicates that 244 is an outlier.

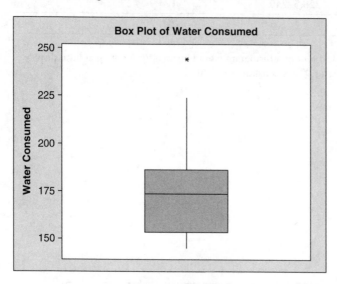

70. (a) **Stem-and-Leaf Display: Wingspans**
Stem-and-leaf of Wingspans N = 22
Leaf Unit = 1.0

```
   3    2 899
   9    3 012233
  (10)  3 5677788999
   3    4 01
   1    4 6
```

(b) With the absence of outliers, the mean—35.4 mm—is an appropriate measure of central tendency. The median—36.5 mm—is also an appropriate measure of central tendency.

(c) The box plot indicates that there are no outliers. There are no data points in either group that are more than 1.5(*IQR*) above Q3 or more than 1.5(*IQR*) below Q1.

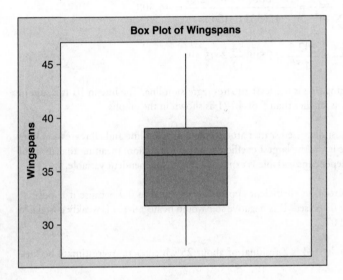

(d) The third quartile is the same as the 75th percentile. 39 is the wingspan for the 75th percentile.

Chapter 3: Two-Variable Data Analysis

71. (C) The data points are trending linearly with a positive slope.

72. (C) The more time spent studying is related to a higher test scores since there is a positive linear relationship.

73. (B) $\hat{y} = 24.765 + 9.051x$. There is positive linear correlation.

$$m = \frac{n\sum xy - \sum x \sum y}{n\sum x^2 - (\sum x)^2} = \frac{13(4617.5) - 60(865)}{13(346) - (60)^2} = 9.051$$

$$b = \frac{\sum y}{n} - m\frac{\sum x}{n} = \frac{865}{13} - 9.051\frac{60}{13} = 24.765$$

74. (D) $\hat{y} = 95.494 - 4.508x$. There is negative linear correlation.

$$m = \frac{n\sum xy - \sum x \sum y}{n\sum x^2 - (\sum x)^2} = \frac{12(3660) - 54(902.5)}{12(332) - (54)^2} = -4.508$$

$$b = \frac{\sum y}{n} - m\frac{\sum x}{n} = \frac{902.5}{12} + 4.508\frac{54}{12} = 95.494$$

75. (D) The best-fitting line is the least-squares regression line. The line in III is a sum of squares of 3.6, which is smaller than 5 or 4.171 as shown in the graphs.

76. (C) III has the least dispersed scatter around the regression line and all regression lines are positive; therefore it has the largest coefficient of determination, meaning that more of the variation of the dependent variable is explained by the independent variable.

77. (D) III has a correlation coefficient approximately equal to 0.9 because it is positive and the scatter is least dispersed. II is negative so r would be negative; I is weakly correlated so r would be around 0.5.

78. (E) The point indicated is a residual of about 25 which is an overestimate because $25 \approx y - \hat{y} > 0$.

79. (B) The point is influential; therefore r decreases and slope decreases. Since this point is an outlier, r is reduced; because the outlier is above and to the left of the center of the data, the slope is tilted and flattened (smaller positive value).

80. (D) The appropriate regression would be quadratic since the points are in a parabolic shape. Taking the square root of each data point y-value would be the appropriate transformation in order to perform a linear regression.

81. (D) The scatter plot of the data shows that the data are clearly nonlinear.

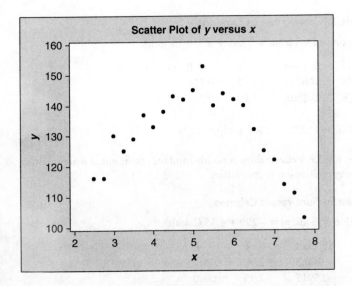

The data point y-value would need to be transformed to achieve linearity and taking the square root would be the appropriate transformation.

82. (B) The scatter plot and coefficient of determination indicate that earnings and dividends are unrelated.

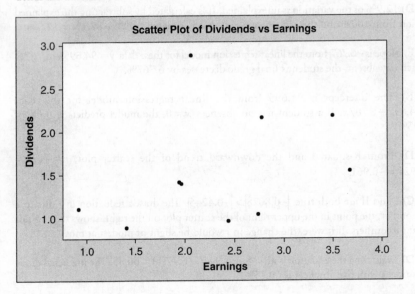

Regression Analysis: Dividends versus Earnings

The regression equation is: Dividends = 0.949 + 0.258 Earnings

Predictor	Coef	SE Coef	T	P
Constant	0.9493	0.6233	1.52	0.166
Earnings	0.2578	0.2508	1.03	0.334

$S = 0.651372$ R-Sq = 11.7%   R-Sq(adj) = 0.6%

83. (B) $\hat{y} = -299.5 + 4.347x$. Because slope is positive and the coefficient of determination is 0.871, there is strong positive linear correlation.

Regression Analysis: Sodium versus Calories

The regression equation is: Sodium = $-$ 299 + 4.35 Calories

Predictor	Coef	SE Coef	T	P
Constant	−299.5	100.3	−2.99	0.017
Calories	4.3469	0.5917	7.35	0.000

$S = 32.6453$ R-Sq = 87.1%   R-Sq(adj) = 85.5%

84. (C) R-squared is a measure of the explained variation in the response variable by the predictor variable.

85. (D) 12.9% of the variation is unexplained. It is calculated by subtracting the explained variation from 100% and the explained variation is R-squared, 100% − 87.1% = 12.9%.

86. (E) Slope is −6.767 from the linear regression model for these data $\hat{y} = 94.695 - 6.767x$. For each day absent, the student's final grade decreases by 6.767%.

87. (B) The intercept is 94.695 from the linear regression model for this data $\hat{y} = 94.695 - 6.767x$. If a student has no absences $x = 0$, the model predicts a grade of 94.695%.

88. (D) From R-squared and the downward trend of the scatter plot, $r = -\sqrt{r^2} = -\sqrt{0.924} = -0.961$.

89. (C) I and II are both true. $|-0.8078| > |-0.4245|$. The drastic reduction in r due to a single outlier, the point in the upper right of the scatter plot on the right, shows that r is not resistant to outliers. If it were, the change in r would be slight or modest at most.

90. (D) Evaluating the equation with = 5, $\hat{y} = 94.7 - 6.77(5) = 60.85$. The predicted grade for a student with five absences is 60.85%.

91. (B) The model is $\hat{y} = 64.833 - 1.754x$. When $x = 19$, $\hat{y} = 64.833 - 1.754$ (19) = 31.507 or 31.5 when rounded to the same precision as the dataset.

92. (C) In Regression 1, R-sq, the explained variation by the predictor variable, is 81.5% while in Regression 2 the R-Sq is 80.7%. More of the variation in the response variable y is explained by x in Regression 1.

93. (D) $r = -0.903$. This indicates strong negative linear correlation.

94. (E) The regression model is $\hat{y} = 64.833 - 1.754x$. Slope is -1.754, which means that for each unit of x, y decreases by 1.754.

95. (C) $r = 0.488$ and is a measure of the strength and direction of a linear relationship; therefore it indicates a weak positive linear relationship. The scatter plot indicates that a transformation would be appropriate before fitting a linear model to these data.

96. (C) The residual pattern indicates that a line would fit after transforming each data point y-value by taking the square root. Further, if this data is regressed without a transformation, the correlation coefficient will indicate a weak linear relationship. However, we cannot tell whether it will be positive or negative from this scatter plot.

97. (C) The residuals lack a pattern, which indicates that a line would fit. Further, if these data are regressed, the correlation coefficient will indicate a strong linear relationship because nearly all the residuals are less than 10% of their corresponding fitted values. However, we cannot tell whether it will be positive or negative from this scatter plot.

98. (B) The residual pattern indicates that a line would fit after transforming each data point y-value by taking the logarithm or the square root of the y-values. Further, if these data are regressed without a transformation, the correlation coefficient will indicate a weak linear relationship. However, we cannot tell whether it will be positive or negative from this scatter plot.

99. (E) The residuals in a regression are also known as errors, so the sum of squared errors (SSE) is listed in the table by residual error under the SS column heading, $SSE = 197.3$. SSE is the unexplained variation in the response variable.

100. (C) The coefficients of determination are 92.4% and 48.6%, respectively. That shows that absences explain more of the variation in the final then the midterm. Because the regression is a least-squares linear regression, the sum of the residuals is always zero, the same for both.

101. (E) $r^2 = \dfrac{SST - SSE}{SST} = \dfrac{2.8185}{3.9140} = 0.720$ or 72.0%. The coefficient of determination is a measure of the variation of the dependent (response) variable that is explained by the regression line and the independent (explanatory) variable.

102. (E) The percentage of unexplained variation in the dependent (response) variable is
$100\% - r^2\% = 100\% - 1.8\% = 98.2\% \cdot r^2 = \dfrac{SST - SSE}{SST} = \dfrac{76.4}{4196.9} = 0.018.$

103. (D) $r = -0.730$. The best explanation for the relationship is that students expecting higher grades study fewer hours.

104. (C) I and II only. Of the two regressions, grade to date on points attempted has a greater r^2; therefore more of the variation in grade to date is explained by points attempted than by online activity. The sum of the residuals is always zero regardless, making it the same for both.

105. (E) The regression equation is $\hat{y} = -0.0026 + 0.000373x$. When $x = 900$, $\hat{y} = -0.0026 + 0.000373(900) = 33.31\%$.

106. (B) The slope is 0.000373. For every point attempted, the grade to date increases by 0.0373%.

107. (A) The intercept is -0.0026. If you don't attempt a single point, the model predicts a grade to date of -0.26% or virtually 0%.

108. (A) $r = \sqrt{.907} = 0.952$. This is a strong positive linear relationship.

109. (C) I and II only. $0.824^2 > 0.798^2$. Additionally, the regression from 1 explains 67.9% of the variation in y while 2 explains 63.7% of the variation in y.

110. (A) The regression equation is $\hat{y} = -0.0026 + 0.000373x$. When $x = 3000$, $\hat{y} = -0.0026 + 0.000373(3000) = 33.31\%$.

111. (D) $\hat{y} = 9.19$. This is an inappropriate extrapolation since the year 2075 is 73 years into the future.

112. (E) There are three assumptions that should be verified for a valid least-squares regression line. They are that the sample was randomly selected, that variance is constant, and that the residuals are normally distributed.

113. (D) The variance of y increases near the mean of y. In the low and high grades, the variance decreases.

114. (B) The residuals from the first dataset with Brain 1 appear to be quite random. There may be a slight violation of the constant variance assumption, but a linear regression will fit quite well. The residuals from the second dataset with Brain 2 appear to have a parabolic shape, which indicates that the data need to be transformed before a linear regression will be a good fit.

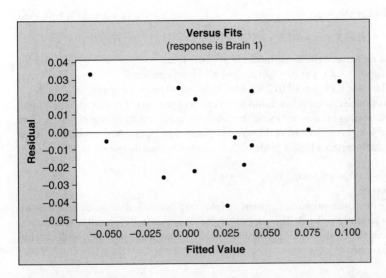

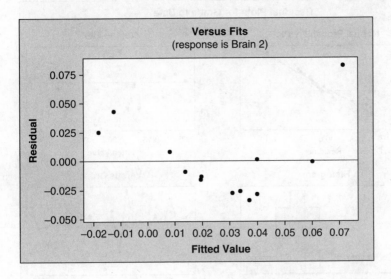

115. (D) Model 1 is the best because the total sum of the squares in Model 1 is 0.030277, which is less than Model 2's sum of 0.034453. Correlation between two variables is the same regardless of the linear model that is fit.

116. (a) Reading two points from the graph on the regression line, (3, 5.8) and (3.5, 5.1), calculate the slope: $m = \dfrac{5.1 - 5.8}{3.5 - 3} = -1.4$. Using $y = -1.4x + b$ and (3, 5.8), then $5.8 = -1.4(3) + b$. $b = 5.8 + 1.4(3) = 10$.

The estimate of the linear model is $\hat{y} = 10 - 1.4x$.

(b) When $x = 3.25$, $\hat{y} = 10 - 1.4(3.25) = 5.45$ hours per week.

(c) When $x = 3.25$, $\hat{y} = 10.012 - 1.416(3.25) = 5.41$ hours per week.

(d) The negative linear relationship between average expected grades and study time is surprising because we think that studying more results in higher grades. This study, however, looked at the *expected* grades, and this seemed to indicate that if students expect a higher grade such as an easy A, they do not study as much.

117. (a) $\hat{y} = -0.0026 + 0.000373x$

(b) $r = 0.952$

(c) The two assumptions of constant variance and normally distributed residuals are not seriously violated. There appears to be an outlier in the plot that, if removed, would rectify the variance violation. See the arrows in the plots below identifying the outlier.

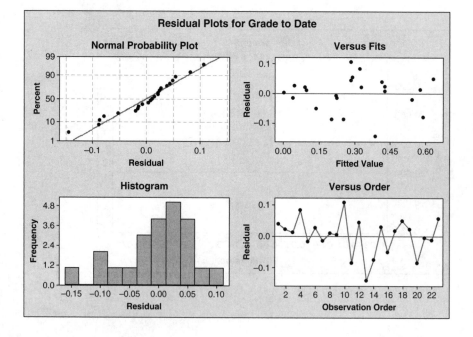

(d) Yes, it does. The more points attempted, the higher the grade.

(e) The scatter plot of grade to date versus online activity does not show a trend. The outlier with 1,200 items completed from online activity in the last 2 weeks is extremely unusual. An explanation could be that students are previewing assignments, quizzes, and tests to get the answers and then retaking them to raise their grades.

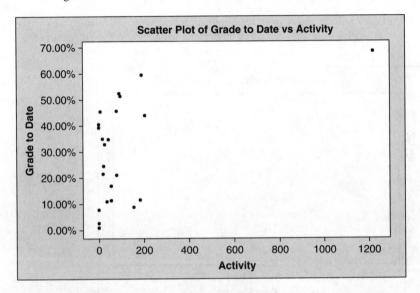

118. (a) The scatter plot does not show any relationship. Calculating the correlation coefficient: $r = -0.293$, which indicates a weak negative linear relationship.

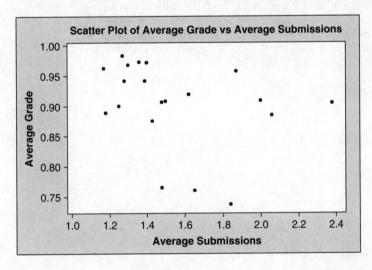

(b) No, the data do not seem to support that. If the instructor's notions were exhibited in the data, there would be an increasing trend.

(c) They would not have enough time to redo and correct the problems they miss to improve their grades. Procrastination might be a better predictor of grades and could also be a predictor of average submissions. That would make procrastination a lurking variable.

119. (a) The scatter plot shows an increasing linear relationship. Calculating the correlation coefficient: $r = 0.869$, which indicates a strong positive linear relationship.

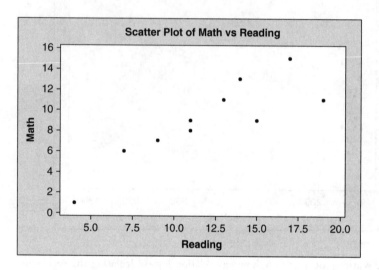

(b) $\hat{y} = 0.06 + 0.745x$. When $x = 12$, $\hat{y} = 0.06 + 0.745(12) = 9$.

(c) The coefficient of determination, r^2, is 0.755, which means that 75.5% of the variation in math scores is explained by reading scores.

(d) Reviewing the residual plots and the scatter plot, it appears that variation is increasing. A researcher might gather a larger sample to see whether this pattern persists. If the pattern persists, transforming the data may help.

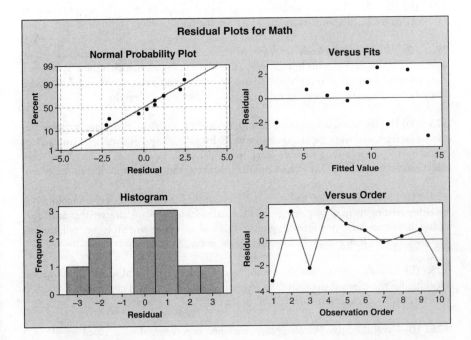

120. **(a)** Reading two points from the graph on the regression line, (1980, 3.35) and (2000, 3.55), calculate slope: $m = \dfrac{3.55 - 3.35}{2000 - 1980} = 0.01$. Using $y = 0.01x + b$ and (1980, 3.35), then $3.35 = 0.01(1980) + b \cdot b = 3.35 - 0.01(1980) = -16.45$. The estimate of the linear model is $\hat{y} = -16.45 + 0.01x$.

(b) When $x = 2016$, $\hat{y} = -16.45 + 0.01(2016) = 3.71$ GPA.

(c) When $x = 2016$, $\hat{y} = -16.451 + 0.010(2016) = 3.709$ GPA. When $x = 2036$, $\hat{y} = -16.451 + 0.010(2036) = 3.909$ GPA. It is surprising that average GPA is approaching a 4.0.

(d) Restrict extrapolation into the future. The model covers a 30-year span but using it to predict more than 5 years into the future is not acceptable.

Chapter 4: Design of a Study: Sampling, Surveys, and Experiments

121. (E) III only. This is a completely randomized design because the treatments were randomized to the experimental units, which are the plots growing the same crop.

122. (E) III only. Completely randomized design controls for lurking variables, which have an effect on the outcome but are not part of the investigation. This is done by random assignment of treatments to experimental units and the assumption that the random effects from the lurking variables will not infuse systematic bias into the results.

123. (D) II only. Block design controls for confounding variables, which are variables that cannot be separated from the treatment variable. By sorting experimental units into groups based on the confounding variable (such as gender), blocks can control for the specific effect gender might have. The effect from the treatment under study can then be measured.

124. (C) I and II only. Matched-pairs design controls for variation in experimental units by either reusing them or by pairing experimental units by similar characteristics and separating them into different treatment groups. It is a block design since it is the confounding variables that are being controlled by this organization.

125. (E) III only. Treatments are randomized to the experimental units. There was no blocking for the effect of the cups. The researchers are aware which cups received the first treatment and which received the second treatment.

126. (B) I and III only. The researchers matched experimental units based on physical characteristics that might influence the effect from the medicine. This is a blocking technique. Additionally, since the participants and the researchers were not informed of the distribution of treatments, this is a double-blind technique used to reduce the placebo effect and experimenter's bias.

127. (A) Statistical methods only identify relationships; however, when used in the controlled environment of experimentation, cause and effect can be supported.

128. (A) The purpose of experimental design is to anticipate and control variation. Block design specifically targets variation from confounding variables by grouping to control that variation.

129. (C) The purpose of experimental design is to anticipate and control variation. Matched-pairs design specifically targets variation from confounding variables. By pairing, variation stemming from a confounding variable can be detected.

130. (A) Since the comparison is between high-protein kibble and a control kibble, matched pairs would be the best design to detect between group differences or effects.

131. (D) II only. Matched-pairs design is the best description since there are only two treatments, i.e., nothing and the oil extract. Each experimental unit receives both treatments.

132. (C) The correct method is to randomly assign the dogs to one of two groups. Then randomly assign the treatments to the groups. This avoids selection and experimenter bias.

133. (D) It appears that gender confounds the results for this appetite-control drug and that blocking by gender would control for the gender difference in response.

134. (B) After the interviews, it is clear that the complicated operation of the new treadmill affected the outcomes and was not part of the investigation. This is a lurking variable.

135. (A) I only. The design of experiments anticipates confounding variables and controls for them by blocking. Replication is used to reduce chance variation and randomization is used to avoid systematic bias.

136. (C) If the difference between what we expect and what we observe is too large to attribute to chance, then it is statistically significant.

137. (D) II only. When the outcomes of an experiment are affected by the participants and researchers knowing who is receiving the real treatment, a double-blind experiment is appropriate and controls for both researcher bias and placebo effect.

138. (B) When an effect is statistically significant, the difference between the expected and observed effect is too large to be attributed only to chance. Hypothesis testing is the method used to determine statistical significance.

139. (D) Observational studies are useful when it is unethical to experiment on subjects because of the adverse effects from treatment levels.

140. (D) II only. An experiment has the advantage that the researcher can control the treatment levels and the environment that subjects are exposed to.

141. (C) This is a stratified sample because the population was first divided into sections and then randomly sampled from each section with the proportion from each block dictated by the population proportions.

142. (E) This is a simple random sample because each member of the population has an equal chance of being chosen and all samples of size 20 have an equal chance of being chosen from the population.

143. (B) This is a systematic sample. Randomization is used to determine a starting position and then the entire population is sampled based on a well-defined pattern—in this case, every 10th attendee.

144. (D) This is a cluster sample. The individual schools are the sections and one of them is selected at random.

145. (A) This is a random sample because each of the three representatives from a course was selected randomly. However, the final sample is not a simple random sample because not all samples of size 30 have an equal chance of being formed in this manner.

146. (B) This is a convenience sample because the friends that you have are easily sampled and you have decided to use them since they appear representative to you. The danger in this type of sampling is that a convenient sample is rarely a representative sample.

147. (E) This is a self-selected or voluntary response sample. The danger in this type of sampling is non-response bias.

148. (C) This is a quota sample. The registration proportions are maintained but the attendees are selected according to their characteristic as retirees.

149. (E) III only. The purpose of sampling is to gather a representative group.

150. (A) Voluntary response bias due to the fact that the respondents had strong feelings about the issue, which motivated them to respond. This resulted in a large number of responses.

151. (B) Wording bias is illustrated here. If the question had been rewritten to emphasize the benefits of the bond for local children, the response would likely have been more favorable.

152. (E) Non-response bias, which could be linked to the time of day the survey is conducted or to a lack of interest.

153. (C) Response bias: due to the stereotypes held about sexuality, men and women both may have misrepresented their actual experience to the interviewer. For example, men may have overestimated while women may have underestimated.

154. (B) I and III only. In experiments, researchers impose treatments on experimental units or subjects in order to manipulate a response that can be linked to the treatment. In fact, the measured response can be called the effect of the treatment.

155. (D) II only. In observational studies, the researcher simply observes and records behavior. There is no attempt to impose a treatment to manipulate a response.

156. (A) I only. The placebo effect is when a treated person shows a benefit despite the fact that the treatment has no effective component.

157. (B) I and III only. When neither the subjects nor the researchers know which groups are receiving treatments or placebos, the experiment is double-blind.

158. (A) I only. An experiment employs statistical control, which refers to a researcher holding constant variables that aren't under study but could influence the outcomes.

159. (A) The appropriate method is to use a table of random digits or a random number generator. Other methods are not considered random due to human bias.

160. (E) III only. In blocking, treatments are randomized within blocks.

161. (A) When the difference between what we expect to find without treatment and what we find with treatment is too large to attribute to chance, then the difference is statistically significant.

162. (A) This is an observational study because the subjects did not receive a treatment. They were simply measured for weight and metabolic rate.

163. (B) When neither the subjects nor the researchers know who is receiving a treatment or a placebo, the experiment is double-blind.

164. (B) I and III only. This is an experiment since the researcher imposed treatments on the groups. It is also a matched-pairs design because each subject was measured before and after for rage level.

165. (A) I only. Blocking should be done on gender to control for the variability due to differences between males and females. Diet and exercise are the treatments.

166. (a) There are $4 \times 4 = 16$ different treatments from the combination of the two fertilizers:

	Levels	0	Phosphate 20	40	60
	0	0,0	0,20	0,40	0,60
Nitrogen	20	20,0	20,20	20,40	20,60
	40	40,0	40,20	40,40	40,60
	60	60,0	60,20	60,40	60,60

With 64 similar plots, randomly assign each treatment combination of fertilizers to four plots since $64 \div 16 = 4$.

(b) Divide each field into 16 plots and randomly assign each treatment combination of fertilizers to the plots field so that each treatment is used four times, once in each field.

(c) The researcher would wish to see an effect of increased yield from the fertilizer. Ideally, he or she would like to attribute the differences between the groups to the fertilizer treatment only and conclude that the fertilizer causes the increase in yield.

167. (a) Randomly assign 10 subjects to each of the four treatment levels.

(b) Randomly assign five males and five females to each of the four treatment levels.

(c) This is a matched-pairs design with a before and after measurement of the rage index.

(d) The lurking variable that could have an effect on the outcomes is experience with technology or teaching experience. Participants could be pre-assessed for levels of expertise in technology and teaching experience and grouped to block for variation just like blocking for gender above.

168. (a) Randomly assign the females to one of the four treatments—music, meditation, exercise, and the new pain medication—so there are 12 subjects in a group. Run the experiment and measure pain levels after the treatments are complete.

(b) Make a control group that receives a placebo to measure the psychological effect of receiving a treatment.

(c) Keep the treatment group of each subject hidden from the researchers who are doing the pain level screenings.

169. (a) The number of absences during the semester and the grade points earned in that class contributing to GPA.

(b) This is an observational study, which can determine that the variables are related but cannot be used to determine cause and effect.

(c) The college could have half the teachers implement group work in the classes while the other half maintains traditional instruction as a control group.

(d) Attendance is confounded with other variables, such as income, illness, family responsibility, and risky behavior.

170. (a) This is an observational study, which can determine that the variables are related but cannot be used to determine cause and effect.

(b) Divide the AD children into two groups. One group will receive the treatment and the other group a pill that looks exactly like the treatment pill. The group receiving the placebo can be used to control for the placebo effect.

(c) To make the study double-blind neither the subjects nor the researchers can know who is receiving the treatment or the placebo. This is done to control for bias on the part of researchers in their evaluation of the outcomes. It also controls for the placebo effect.

Chapter 5: Probability and Random Variables

171. (A) A chance experiment or random phenomenon can be observed and measured but cannot be predicted. An example would be rolling a fair die.

172. (A) Individual outcomes are termed simple events.

173. (E) Sample space is the set of all possible outcomes, or simple events, of a random experiment.

174. (D) I and II only. The axioms of probability stipulate that $P(E) = 0$ if the event will not happen, $P(E) = 1$ if the event is certain to happen, and between 0 and 1 to express the fraction of time the event will happen. Furthermore, the sum of all the probabilities of all possible outcomes must be exactly 1.

175. (D) The $P(Male) = P(Female) = \dfrac{1}{2}$. There are 10 ways to form all the possible litters. The probability is $10\left(\dfrac{1}{2}\right)^2\left(\dfrac{1}{2}\right)^3 = \dfrac{5}{16}$.

176. (D) The $P(Male) = P(Female) = \dfrac{1}{2}$. There are 10 ways to form all the possible litters of two males and three females and 10 ways to form all possible litters of three males and two females. The probability is $10\left(\dfrac{1}{2}\right)^2\left(\dfrac{1}{2}\right)^3 + 10\left(\dfrac{1}{2}\right)^3\left(\dfrac{1}{2}\right)^2 = \dfrac{5}{8}$.

177. (A) The $P(Male) = P(Female) = \dfrac{1}{2}$. There are 10 ways to form all the possible litters of two males and three females and 10 ways to form all possible litters of three males and two females. The probability is $10\left(\dfrac{1}{2}\right)^2\left(\dfrac{1}{2}\right)^3 + 10\left(\dfrac{1}{2}\right)^3\left(\dfrac{1}{2}\right)^2 = \dfrac{5}{8}$. The probability of not getting either of these two litters is the complement, $1 - \dfrac{5}{8} = \dfrac{3}{8}$.

178. (D) This is the probability of drawing a queen (Event A) given that you have drawn two kings and two queens (Event B) already. To calculate $P(A|B)$, consider the reduced configuration of the card deck. There are only two queens left in the remaining 48 cards. $P(A|B) = \dfrac{2}{52-4} = \dfrac{1}{24}$.

179. (A) This is the probability of drawing a king or a queen (Event A) given that you have drawn two kings and two queens (Event B) already. To calculate $P(A|B)$, consider the reduced configuration of the card deck. There are only two kings and two queens left in the remaining 48 cards. $P(A|B) = \dfrac{2+2}{52-4} = \dfrac{1}{12}$.

180. (E) The probability distribution for Bill's serving in tennis is:

Event, E	P(E)
Ace	0.15
Double Fault	0.25
Not an Ace or Double Fault	X
Total	1.00

$P(not\ an\ Ace\ or\ Double\ Fault) = 1 - (0.15 + 0.25) = 0.6$

181. (C) II only. Random variables are both discrete and continuous. They are the outcomes of a random experiment.

182. (D) The number of minutes in an hour is 60, a constant.

183. **(E)** Since the probabilities must sum to 1 let x be the unknown value, $1 = \frac{1}{5} + x + \frac{1}{5} + \frac{1}{6} + \frac{1}{6} + \frac{1}{8}$. Solving for x, the probability needs to be $\frac{17}{120}$.

184. **(D)** Let x represent the probability of the odd values. The $2x$ would be the probability of the even values. Since the probabilities must sum to $1, 1 = x + 2x + x + 2x + x + 2x = 9x$; therefore, $x = \frac{1}{9}$ which is the probability of the odd values and the probability of the even values is twice that or $\frac{2}{9}$.

185. **(D)**

$$E(X) = \sum xP(x) = 1(0.4) + 2(0.3) + 3(0.2) + 4(0.05) + 5(0.025) + 6(0.025)$$
$$= 2.075 = 2.1.$$

$$V(X) = \sum (x - \mu_X)^2 P(x) = (1 - 2.075)^2 (0.4) + (2 - 2.075)^2 (0.3) + (3 - 2.075)^2 (0.2)$$
$$+ (4 - 2.075)^2 (0.05) + (5 - 2.075)^2 (0.025) + (6 - 2.075)^2 (0.025) = 1.4194 = 1.4.$$

186. **(C)**

$$E(X) = \sum xP(x) = 0(0.1353) + 1(0.2707) + 2(0.2707) + 3(0.1804) + 4(0.0902)$$
$$+ 5(0.0527) = 1.9776 = 2.0$$

$$V(X) = \sum (x - \mu_X)^2 P(x) = (0 - 1.9776)^2 (0.1353) + (1 - 1.9776)^2 (0.2707)$$
$$+ (2 - 1.9776)^2 (0.2707) + (3 - 1.9776)^2 (0.1804) + (4 - 1.9776)^2 (0.0902)$$
$$+ (5 - 1.9776)^2 (0.0527) = 1.8269 = 1.8.$$

187. **(D)**

$$E(X) = \sum xP(x) = 0(0.16807) + 1(0.36015) + 2(0.3087) + 3(0.1323) + 4(0.02835)$$
$$+ 5(0.00243) = 1.5$$

$$V(X) = \sum (x - \mu_X)^2 P(x) = (0 - 1.5)^2 (0.16807) + (1 - 1.5)^2 (0.36015) + (2 - 1.5)^2 (0.3087)$$
$$+ (3 - 1.5)^2 (0.1323) + (4 - 1.5)^2 (0.02835) + (5 - 1.5)^2 (0.00243) = 1.05$$

$$SD = \sqrt{V(X)} = \sqrt{1.05} = 1.0247 = 1.0$$

$$E(X) = \sum xP(x) = 0(0.3385) + 1(0.4481) + 2(0.1883) + 3(0.0251) = 0.9$$

188. (A)

$$V(X) = \sum (x - \mu_X)^2 \; P(x) = (0-0.9)^2 (0.3385) + (1-0.9)^2 (0.4481)$$
$$+ (2-0.9)^2 (0.1883) + (3-0.9)^2 (0.0251) = 0.6172$$

$$SD = \sqrt{V(X)} = \sqrt{1.6172} = 0.7856 = 0.8$$

189. (E) Reading the heights of the bars in the histogram and noting that the scale of the vertical axis is in percents leads to:

Accidents, x	0	1	2	3
P(X)	0.70	0.15	0.10	0.05

190. (A) The scale of the vertical axis is in percents. The sum of the heights of the bars must equal 100. Let x be the unknown bar height. Then $100 = x + 15 + 5 + 5$ and solving for x, $x = 75$.

191. (C) $P(Minnesota \; \& \; tablet) = 0.05(0.08) = 0.004$

192. (E)

$$P(California \,|\, Cell \; phone) = \frac{0.45(0.56)}{0.25(0.78) + 0.45(0.56) + 0.25(0.86) + 0.05(0.92)} = 0.356$$

193. (C)

$$P(California \; or \; Arizona \,|\, Cell \; phone)$$
$$= \frac{0.45(0.56) + 0.25(0.78)}{0.25(0.78) + 0.45(0.56) + 0.25(0.86) + 0.05(0.92)} = 0.631$$

194. (E) Drawing a jack or a heart is not mutually exclusive because there is one jack that is also a heart; therefore $P(Jack \; or \; Heart) = \dfrac{4 + 13 - 1}{52} = \dfrac{4}{13}$.

195. (C) Drawing a jack and a three are mutually exclusive; therefore

$$P(Jack \; and \; 3) = \frac{4}{52}\left(\frac{4}{51}\right) = \frac{16}{663}.$$

196. (B) Since you both choose the same number, there is only one success out of the 20 numbers to choose from. $P(s) = \dfrac{s}{s+f} = \dfrac{1}{20} = 0.05$.

197. (A) $P(4) = \dfrac{6+6-1}{36} = \dfrac{11}{36}$. Think of having a white die and a black die so that you can tell them apart. A six on the white die can turn up with all six of the numbers on the black die and vice versa; however, when you get a double six, that is counted in both.

198. (C) $P(4) = \dfrac{4}{4(5)} = \dfrac{1}{5}$

199. (A) $P(2 \text{ or } 9) = \dfrac{2}{4(5)} = \dfrac{1}{10}$

200. (B) $P(3) = \dfrac{4+5-1}{4(5)} = \dfrac{2}{5}$

201. (C) I and III only. The probability distribution is modeled by a smooth curve, which can be defined by a function. The characteristics of this function are that it is nonnegative since probabilities are nonnegative, and that the area underneath the curve is equal to one since the sum of the probabilities in a distribution must equal one. The probability of an individual event E cannot be calculated by finding $P(X = E)$ since $P(X = E) = 0$.

202. (D) II only. The normal curve is "bellshaped" and symmetric about the mean μ. The area underneath the curve is equal to one, since the sum of the probabilities in a distribution must equal one. Normal distributions do not have to have $\mu = 0$ and $\sigma = 1$. Only the standard normal has $\mu = 0$ and $\sigma = 1$.

203. (E) III only. When it is known that a distribution is approximately normal, the normal distribution is valid for problem solving.

204. (D) $P(z \geq 1.25) = P(z \leq -1.25) = 0.1056$ from the table. This is due to the symmetry of the normal distribution.

205. (D) $P(-2.95 < z \leq 0.95) = P(z \leq 0.95) - P(z < -2.95) = 0.8273$ from the tables.

206. (C) $P(X > 160) = P\left(z > \dfrac{160 - 163.2}{9.3} \approx -0.34\right) = 1 - P(z < -0.34) = 0.6331$ from the tables.

207. (D) $P(X < 180) = P\left(z > \dfrac{180 - 202.3}{50.7} \approx -0.44\right) = P(z < -0.44) = 0.3300$ from the tables.

208. (A) I only. For the man, $P(254.5 < X < 255.5) = P\left(\dfrac{254.5 - 188.3}{66.4} < z < \dfrac{255.5 - 188.3}{66.4}\right) =$

$P(1.00 < z < 1.01) = P(z < 1.01) - P(z < 1.00) = 0.0025$ from the tables. For the woman,

$P(214.5 < X < 215.5) = P\left(\dfrac{214.5 - 155.9}{60.3} < z < \dfrac{215.5 - 155.9}{60.3}\right) = P(0.97 < z < 0.99) =$

$P(z < 0.99) - P(z < 0.97) = 0.0049$ from the tables. A woman weighing 215 pounds is more likely than a man weighing 255 since the probability for the woman is greater than the probability for the man.

209. (E) $P(X > x) = P\left(z > \dfrac{x - 500}{100}\right) = 1 - 0.07 = 0.93$. From the tables, $P(z < 1.48) =$ 0.9306 being conservative in order to make sure that you are in the top 7%. Therefore $1.48 = \dfrac{x - 500}{100}$ and $x = 1.48(100) + 500 = 648$.

210. (A) $P(X > x) = P\left(z > \dfrac{x - 18.5}{4.5}\right) = 0.82$. From the tables, $P(z < 0.92) = 0.8212$ being conservative in order to make sure that 82% of children approximately are below this index value. Therefore $0.92 = \dfrac{x - 18.5}{4.5}$ and $x = 0.92(4.5) + 18.5 = 22.6$.

211. (C) Generally, the greater the number of trials, the higher our confidence is that the relative frequency of successes accurately approximates the true proportion in the population, which is also interpreted as the probability of success.

212. (E) $P(H) = 0.3$ means that 3 out of 10 will be heads; therefore, digits 0, 1, 2 will represent a toss resulting in a head while 3, 4, 5, 6, 7, 8, 9 will represent a toss resulting in a tail.

213. (C) These classes are to be considered independent since the students are all different from class to class. With that assumption, $\mu_{X+Y+Z} = 17 + 16 + 18 = 51$ and $\sigma_{X+Y+Z} = \sqrt{1.25^2 + 1.85^2 + 2.35^2} = 3.24$.

214. (E) III only. Adding two random variables, $X + Y$, results in a new distribution with a mean that is the sum of the means of X and Y. The variance of the new distribution is the sum of the two variances only if the two variables are independent. If X and Y are dependent, the variances are summed along with an adjustment for the correlation, $2\rho\sigma_X\sigma_Y$, where ρ is the correlation.

215. (D) I and II only. Adding a constant a to a random variable will change the mean by that constant, $a + \mu_X$, but the variance will remain unchanged. This is simply a shift of the entire distribution up a units. Multiplying a random variable by a constant b will change the mean proportionally, $b\mu_X$. It is either an enlargement or a reduction of the distribution. This has the effect of changing the variance by the square of b, $b^2\sigma_X^2$.

216. (D) The described situation represents a conditional probability since we know one property already (the fact that the worker is full-time). Since we know the worker is full-time, this statement should be on the right-hand side of the conditional probability statement.

217. (C) Since this is a discrete random variable, the probability can be found using $P(X = 1) + P(X = 2) + P(X = 3)$.

218. (A) When two events are independent, the "and" probability involving both events can be found by multiplying the individual probabilities.

219. (D) P(male or takes fewer than two trips per year) $= P$(male) $+ P$(takes fewer than two trips per year) $- P$(male and takes fewer than two trips per year) $= 0.50 + 0.36 - 0.17$.

220. (a) The probability distribution is found by finding the total (1,227) and dividing each frequency by the total. Note that the probabilities should add to approximately 1 (may not be exact depending on rounding).

X	1	2	3	4	5	6	7	8	9
P(X)	0.13	0.18	0.31	0.15	0.10	0.07	0.05	0.01	0.003

(b) The question is asking for the expected value, which is also known as the mean of the probability distribution. Using the formula $1(0.13) + 2(0.18) + 3(0.31) + 4(0.15) + 5(0.10) + 6(0.07) + 7(0.05) + 8(0.01) + 9(0.003) = 3.397$ we would expect someone to throw between three and four times before missing.

221. (a) Complete the marginal column and marginal row with totals, and then calculate the cell probabilities by dividing each frequency by the total number of donors—290. The tables below show the complete distribution.

	Blood Type				
Rh-factor	O	A	B	AB	Row Totals
Positive	110	98	26	9	243
Negative	20	18	6	3	47
Column Totals	130	116	32	12	290

Probabilities calculated to four decimals.

	Blood Type				
Rh-factor	O	A	B	AB	Row Totals
Positive	0.3793	0.3379	0.0897	0.0310	0.8379
Negative	0.0690	0.0621	0.0207	0.0103	0.1621
Column Totals	0.4483	0.4000	0.1103	0.0414	1.0000

(b) $P(type\ A) = \dfrac{116}{290} = 0.3379 + 0.0621 = 0.4000$

$P(Rh\ positive) = \dfrac{243}{290} = 0.3793 + 0.3379 + 0.0897 + 0.0310 = 0.8379$

$P(not\ type\ O) = 1 - \dfrac{130}{290} = 1 - 0.4483 = 0.5517$

$P(type\ O\ or\ type\ AB) = \dfrac{130}{290} + \dfrac{12}{290} = \dfrac{142}{290} = 0.4483 + 0.0414 = 0.4897$

$P(type\ O\ or\ Rh\ negative) = \dfrac{130}{290} + \dfrac{47}{290} - \dfrac{20}{290} = \dfrac{157}{290} = 0.4483 + 0.1621 - 0.0690 = 0.5414$

(c) $P(type\ O\ negative) = 0.0690$. The number of donors they would expect is $467 \cdot P(type\ O\ negative) = 467(0.0690) = 32$.

222. (a) The following is a proposed assignment of digits to colors. Other answers may also be correct.

Digit	Color
0	Yellow
1	Green
2	Red
3	Brown
4	Blue
5	Orange
Digits 6–9	No assignment. SKIP.

(b) Reading across by rows for 30 trials results in the following tabulation.

Digit	Color	Frequency
0	Yellow	8
1	Green	4
2	Red	6
3	Brown	3
4	Blue	3
5	Orange	6
Digits 6–9	No assignment. SKIP.	25

(c) Calculating empirical probabilities using $P(E) = \dfrac{f}{n}$ where E represents a color.

Digit	Color	Frequency	P(E)
0	Yellow	8	4/15
1	Green	4	2/15
2	Red	6	1/5
3	Brown	3	1/10
4	Blue	3	1/10
5	Orange	6	1/5
Digits 6–9	No assignment. SKIP.	25	

Classical probability based on a uniform distribution means that the colors are equally likely.

Color	P(E)
Yellow	1/6
Green	1/6
Red	1/6
Brown	1/6
Blue	1/6
Orange	1/6

(d) The law of large numbers states that the proportion of successes in the simulation approaches the classical (theoretical) probability with a very large number of trials.

223. (a) $P(805.0 < w < 1195.0) = P(w < 1195.0) - P(w < 805.0) = 0.9829$

Standardizing and using tables:

$$P(w < 1195.0) - P(w < 805.0) = P\left(z < \frac{1195.0 - 1005.0}{81.6} = 2.33\right)$$

$$-P\left(w < \frac{805.0 - 1005.0}{81.6} = -2.45\right) = 0.9901 - 0.0071 = 0.9830$$

(b) $P(l > 405.0 \text{ or } l < 375.0) = P(l > 405.0) + P(l < 375.0) = 0.5008$

Standardizing and using tables:

$$P(l > 405.0) + P(l < 375.0) = P\left(z > \frac{405.0 - 385.0}{21.7} = 0.92\right)$$

$$+ P\left(z < \frac{375.0 - 385.0}{21.7} = -0.46\right) = 0.1788 + 0.3228 = 0.5016$$

(c) $P(404.5 < l < 405.5) = 0.0120$

Standardizing and using tables:

$$P(l < 405.5) - P(l < 404.5) = P\left(z < \frac{405.5 - 385.0}{21.7} = 0.94\right)$$

$$- P\left(z < \frac{404.5 - 385.0}{21.7} = 0.90\right) = 0.8264 - 0.8159 = 0.0105$$

$$P(1194.5 < w < 1195.5) = 0.0003$$

Standardizing and using tables:

$$P(w < 1195.5) - P(w < 1194.5) = P\left(z < \frac{1195.5 - 1005.0}{81.6} = 2.33\right)$$
$$- P\left(w < \frac{1194.5 - 1005.0}{81.6} = 2.32\right) = 0.9901 - 0.9898 = 0.0003$$

Since $P(404.5 < l < 405.5) = 0.0105 > P(1194.5 < w < 1195.5) = 0.0003$, it is more likely to get a greater spot-nosed monkey that is 405 mm in length than 1195 g in weight.

224. (a) $P(X < 500) = P\left(z < \frac{500 - 515}{116}\right) = 0.449$. The score of 500 would be the 45th percentile.

$$P(X < 720) = P\left(z < \frac{720 - 515}{116}\right) = 0.961.$$ The score of 720 would be the 96th percentile.

(b) $$P(X < 500) = P\left(z < \frac{500 - 511}{136}\right) = 0.468.$$ The score of 500 would be the 47th percentile.

$P(X < 720) = P\left(z < \dfrac{720 - 511}{136}\right) = 0.938$. The score of 720 would be the 94th percentile.

(c) For 2009, $P(X < x) = P\left(z < \dfrac{x - 515}{116}\right) = 0.90$. The score of 664 would be the 90th percentile.

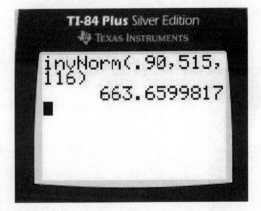

For 1999, $P(X < x) = P\left(z < \dfrac{x - 511}{136}\right) = 0.90$. The score of 685 would be the 90th percentile.

Year 1999 had higher scores above the 90th percentile since the score of 685 is higher than the 2009 score of 664.

For 2009, $P(X < x) = P\left(z < \dfrac{x - 515}{116}\right) = 0.25$. The score of 437 would be the 25th percentile.

For 1999, $P(X < x) = P\left(z < \dfrac{x-511}{136}\right) = 0.25.$ The score of 419 would be the 25th percentile.

Year 2009 had higher scores below the 25th percentile since the score of 437 is higher than the 2009 score of 419.

(d) I would agree with the news article. The average for year 2009 is a little higher than 1999, but the standard deviation is larger in 1999. The analysis of the percentile rankings shows that the lowest score in the top 10% of the scores in 1999 was higher by 21 points. Further, the highest score in the bottom 25% of the scores in 1999 was 18 points lower than scores in 2009. With the normal distribution assumption, in 2009 34% of the students scored between 515 and 631 while in 1999 34% of the students scored between 511 and 647. This demonstrates that above the first standard deviation and due to a greater standard deviation, there are more scores that are higher in 1999.

225. (a) Course #1: $\bar{X} = 17.2$ and $s = 1.9.$ Course #2: $\bar{Y} = 15.5$ and $s = 1.4.$

(b) $P(X < x) = P\left(z < \dfrac{X-17.2}{1.9} = 1.28\right) = 0.9.$ $X = 19.6.$

$P(Y < y) = P\left(z < \dfrac{X-15.5}{1.4} = 1.28\right) = 0.9.$ $X = 17.3$.

It appears that Course #2 has a lower score for its 90th percentile which would indicate that it has a higher level of difficulty. This means that Course #1 appears to be easier.

(c) For the total score $X + Y$, the mean is $\bar{X} + \bar{Y} = 17.2 + 15.5 = 32.7.$ The standard deviation is $\sqrt{s_X^2 + s_Y^2} = \sqrt{1.9^2 + 1.4^2} = 2.4.$

Chapter 6: Binomial Distribution, Geometric Distribution, and Sampling

226. (B) The binomial distribution has mean np and standard deviation $\sqrt{np(1-p)}$ where p is the probability of success. Since the probability of failure is 0.29, the probability of success is 0.71. In this case the mean is $145(0.71) = 102.95$ and the standard deviation is $\sqrt{145(.71)(.29)} = 5.46$.

227. (B) The binomial distribution has mean np and standard deviation $\sqrt{np(1-p)}$ where p is the probability of success. In this case the mean is $75(0.3)$ and the standard deviation is $\sqrt{75(.3)(.7)}$.

228. (A) The binomial distribution has mean np and standard deviation $\sqrt{np(1-p)}$ where p is the probability of success. In this case the mean is $150(0.3)$ and the standard deviation is $\sqrt{150(.3)(.7)}$.

229. (C) The binomial distribution has mean np and standard deviation $\sqrt{np(1-p)}$ where p is the probability of success. Since the probability of failure is 0.4, the probability of success is 0.6. In this case the mean is $50(0.6)$ and the standard deviation is $\sqrt{50(.4)(.6)}$.

230. (A) A random variable which counts successes in independent trials is binomial with mean np: $15(0.3) = 4.5$.

231. (D) A random variable which counts successes in independent trials is binomial with standard deviation $\sqrt{np(1-p)} = \sqrt{20(0.4)(0.6)}$.

232. (A) The mean of a binomial random variable is np. Since there are 50 trials, you can solve for p algebraically: $50p = 20$, $p = 0.4$.

233. (B) The mean of a binomial random variable is np. Since there are 40 trials, you can solve for p algebraically: $40p = 30$, $p = 0.75$. The probability of failure is $1 - p = 1 - 0.75 = 0.25$.

234. (A) The described random variable is binomial. In general, $p(X = x) = \binom{n}{x} p^x (1-p)^{n-x}$.

235. (B) The described random variable is binomial. In general, $p(X = x) = \binom{n}{x} p^x (1-p)^{n-x}$.

236. (A) For a binomial random variable, $p(X = x) = \binom{n}{x} p^x (1-p)^{n-x}$.

237. (A) The described experiment is binomial and the probability of at least 80 heads can be written as $P(X \geq 80) = 1 - P(X < 80)$.

238. (D) The described experiment is binomial and the probability of at most 80 heads is written mathematically as $P(X \leq 80)$.

239. (A) The described experiment is binomial with a success represented as rolling a five. The probability of at most two fives is written mathematically as $P(X \leq 2)$.

240. (B) The described experiment is binomial with a success represented as rolling a three. The probability of more than one three is written mathematically as $P(X > 1) = 1 - P(X \leq 1)$.

241. (C) The described situation is binomial with $n = 300$ and $p = 0.65$. Since np is larger than 10, the normal approximation to the binomial distribution is appropriate with

$$p(X < 200) = p\left(Z < \frac{200 - (300)(0.65)}{\sqrt{300(0.65)(0.35)}}\right).$$

242. (C) The described situation is binomial with $n = 80$ and $p = 1/4$. Since np is larger than 10, the normal approximation to the binomial distribution is appropriate with

$$p(X > 50) = p\left(Z > \frac{50 - (80)(0.25)}{\sqrt{80(0.25)(0.75)}}\right).$$

243. (D) The geometric distribution will give us the probability of success occurring on the nth trial of a binomial experiment. For the 15th trial, $P(X = 15) = 0.3(1 - 0.3)^{15-1}$.

244. (D) The geometric distribution will give us the probability of success occurring on the nth trial of a binomial experiment. Since the probability of success is $1 -$ (probability of failure) we have $P(X = 15) = 0.4(1 - 0.4)^{15-1}$.

245. (C) Since np is larger than 10, the sampling distribution for the sample proportion will be normal with mean p and standard deviation $\sqrt{\dfrac{p(1-p)}{n}}$.

246. (C) If the probability of failure is 0.2, then the probability of success is $1 - 0.2 = 0.8$. Therefore $p(X = 10) = \dbinom{60}{10}(0.8)^{10}(0.2)^{50}$.

247. (B) The binomial distribution has a mean np. In this case $np = 15 = 100p$. Solving this algebraically we find $p = 0.15$ and $p(X = 21) = \dbinom{100}{21}(0.15)^{21}(0.85)^{79}$.

248. (C) This is actually a binomial experiment with 20 trials and a probability of success of $\dfrac{1}{5}$. To receive a 60%, the student must correctly guess the answer on $0.6(20) = 12$ questions. This is equivalent to finding the probability of 12 successes in 20 trials: $P(X = 12) = \dbinom{20}{12}\left(\dfrac{1}{5}\right)^{12}\left(\dfrac{4}{5}\right)^{8}$.

249. **(E)** The described random variable is binomial. Finding the probability of exactly five successes is equivalent to finding $P(X = 5) = \binom{16}{5}(0.2)^5(0.8)^{11} = 0.12$.

250. **(B)** The described random variable is binomial. However, we are given the probability of failure instead of success, which would be $1 - 0.42 = 0.58$. Therefore the probability of exactly eight successes would be $P(X = 8) = \binom{20}{8}(0.58)^8(0.42)^{12} = 0.05$.

251. **(E)** The probability of success for each trial of a binomial experiment is the same. If the probability of success for the fifth trial is 0.8, then it is 0.8 for every other trial, including the sixth.

252. **(A)** The probability of success and failure is the same for each trial of a binomial experiment. If the probability of failure on the third trial is 0.3, then the probability of success is 0.7 for all trials, including the fourth.

253. **(C)** The geometric random variable x counts the number of trials until the first success occurs and $P(X = x) = p(1 - p)^{x-1}$. Therefore $p(1 - p) = 0.25$. This can be solved for the probability of success: $p = 0.5$, and $P(X = 7) = \binom{12}{7}(0.5)^7(0.5)^5 = 0.19$.

254. **(C)** Since the probability of success on each trial is p for a binomial experiment, $p^2 = 0.09$. Solving for p, we have a probability of success of 0.3 and $P(X = 3) = \binom{10}{3}(0.3)^3(0.7)^7$.

255. **(A)** Since the sample size is large ($n \geq 30$), the sampling distribution will be approximately normal with $\mu_{\bar{x}} = \mu = 15$ and $\sigma_{\bar{x}} = \dfrac{\sigma}{\sqrt{n}} = \dfrac{2.1}{\sqrt{35}}$. Thus $P(\bar{x} > 14) = P\left(z > \dfrac{14-15}{2.1/\sqrt{35}}\right)$.

256. **(E)** A large sample will lead to a sampling distribution that is approximately normal with $\mu_{\bar{x}} = \mu = 10$ and $\sigma_{\bar{x}} = \dfrac{\sigma}{\sqrt{n}} = \dfrac{2.5}{\sqrt{45}} = 0.37$.

257. **(E)** Since the sample is small ($n < 30$), the sampling distribution will have a shape similar to the shape of the population, which is unknown.

258. **(D)** The sampling distribution actually consists of the sample means from all possible samples of a certain size. It is the size of the sample, which helps determine the shape of the sampling distribution.

259. **(A)** With a small sample size and a population distribution of unknown shape, the shape of the sampling distribution will also be unknown. However, $\mu_{\bar{x}} = \mu = 10$ and $\sigma_{\bar{x}} = \dfrac{\sigma}{\sqrt{n}}$.

260. **(A)** The geometric random variable x counts the number of trials until the first success and $P(X = x) = p(1 - p)^{x-1} = 0.37(0.63)^3$.

261. (E) For a geometric random variable the probability of the first success occurring on the nth trial is $P(X = n) = p(1 - p)^{n-1}$. Additionally, since x is discrete, the probability that x is 10 or 11 is $P(X = 10) + p(X = 11) = 0.7(0.3)^9 + 0.7(0.3)^{10}$.

262. (A) The probability of failure on any trial is 0.42, therefore the probability of success on any trial is $1 - 0.42 = 0.58$. Since x is geometric, the probability of the first success occurring on the first or second trial is $P(X = 1) + p(X = 2) = 0.58(0.42)^0 + 0.58(0.42)^1$.

263. (A) For a geometric random variable, the average time to the first success is $\dfrac{1}{p} = \dfrac{1}{0.64} = 1.56$.

264. (C) The binomial distribution can be approximated by the normal distribution with $\mu = np$ and $\sigma = \sqrt{np(1-p)}$ when n is large enough. In this case, $P\left(z > \dfrac{25 - 27}{3.29}\right) = P\left(z > \dfrac{x - \mu}{\sigma}\right)$. If $np = 27$ and $n = 45$, $p = 0.6$.

265. (E) Since x is binomial, $P(X = x) = \dbinom{n}{x} p^x (1 - p)^{n-x}$. Therefore $p = 0.45$ and $np = 13.5$, giving us the approximation $P\left(z < \dfrac{x - np}{\sqrt{np(1-p)}}\right) = P\left(z < \dfrac{10 - 13.5}{2.72}\right)$.

266. (D) Since $np \geq 5$, the distribution can be approximated by a normal distribution with $\mu = np = 19.5$ and $\sigma = \sqrt{np(1-p)} = 3.7$.

267. (D) If the probability of failure is 40%, then the probability of success is 60%. Further, since $np \geq 5$, the distribution can be approximated by a normal distribution with $\mu = np = 42$ and $\sigma = \sqrt{np(1-p)} = 4.1$.

268. (C) The described situation is binomcdf with $n = 23$ and $p = 0.17$. The probability that fewer than three trials will be successful is $P(X < 3) = P(X \leq 2) = \dbinom{23}{0}(0.17)^0(0.83)^{23} + \dbinom{23}{1}(0.17)^1(0.83)^{22} + \dbinom{23}{2}(0.17)^2(0.83)^{21} = 0.022$. This can also be found on the calculator using binomcdf $(23, 0.17, 2)$.

269. (A) The described situation is binomcdf with $n = 31$ and $p = 1 - 0.45 = 0.55$. The probability of more than 20 failures is equivalent to 10 or fewer successes, which is $P(X \leq 10)$ symbolically. This is best found using the calculator: binomcdf $(31, 0.55, 10)$.

270. (B) Using the normal approximation to the sampling distribution of $\hat{p}$, this question is asking for $P(\hat{p} > 0.45)$. Since the mean is p and the standard deviation is $\sqrt{\dfrac{p(1-p)}{n}}$ this is equivalent to $P\left(Z > \dfrac{0.45 - 0.55}{\sqrt{\dfrac{0.55(0.45)}{95}}}\right) = P\left(Z > \dfrac{0.45 - 0.55}{0.051}\right)$.

271. (D) $P(X > 1) = 1 - P(X \le 1)$. Using the TI83 or 84 calculator, this can be found with $1-$ binomial $(30, 0.14, 1)$.

272. (E) This probability can be found on the TI83 or 84 calculator using binomial $(12, 0.29, 3)$.

273. (B) The described random variable is binomial, therefore the mean number of successes would be np and the probability of two successes would be found using the binomial formula.

274. (C) If there are fewer than two failures, there must be either one or no failures; in other words either 9 or 10 successes.

275. (a) Each individual probability is found using the formula for exactly that number of successes. For instance, $p(x = 0)$ is the value in the table under 0.

x	0	1	2	3
p(x)	0.216	0.432	0.288	0.064

(b) Since x is discrete, the probability of two or more successes is $p(x = 2) + p(x = 3) =$ $0.288 + 0.064 = 0.352$.

276. (a) Since the sample size is 40, the sampling distribution of the sample mean will be approximately normal with a mean of 10 (assuming the claim is true) and a standard deviation of $\dfrac{1.1}{\sqrt{40}} = 0.1739$. Therefore the probability a sample would have a mean less than 9.5 is $P(\bar{x} < 9.5) = P\left(Z < \dfrac{9.5 - 10}{0.1739}\right) = 0.002$.

(b) Yes, assuming the same sample size. As long as the sample size is larger than 30, the sampling distribution will be approximately normal with a mean of 10 regardless of the distribution of the population. However, the standard deviation depends on the sample size: it decreases as the sample size increases.

277. (a) The distribution will be skewed right (but with less skew than the distribution of the population itself) with a mean of 46.1 and a standard deviation of $\dfrac{5.3}{\sqrt{10}} = 1.67$.

(b) The distribution will be approximately normal with a mean of 46.1 and a standard deviation of $\dfrac{5.3}{\sqrt{50}} = 0.75$.

(c) In both cases the mean will be the same, but the standard deviation decreases as the size of the sample increases. This is true in general for the sampling distribution of sample means.

278. (a) Since the sample size is large, the sampling distribution is approximately normal and $P(\bar{x} \le 115) = P\left(Z < \dfrac{115 - 121.3}{20.9 / \sqrt{40}}\right) = 0.002$.

(b) It is unlikely to get a sample with a mean of 115 or less if the true mean is 121.3. However, such a sample was found. Therefore it is unlikely that the true mean is 121.3 based on the sample from part (a).

279. (a) This is a binomial experiment, and the expected value of x (the number of successes) is $np = 0.82(45) = 36.9$. Therefore you would expect about 37 students to have bought textbooks off campus.

(b) The probability that more than 130 students would have bought textbooks off campus is equivalent to $P(X > 130) = 1 - P(X \le 130) = 1 - $ binomcdf $(150, 0.82, 130) = 0.051$, where the $p = 0.82$ and $n = 150$.

280. (a) Since it is thought that the defects are independent, this can be thought of as a binomial experiment with $n = 150$ and $p = 0.012$. Therefore the $P(X \ge 4) = 1 - P(X < 4) = 1 - $ binomcdf$(150, 0.012, 3) = 0.1075$.

(b) Again, since it is thought that the defects are independent, the probability any one part is defective will be 1.2% regardless of how many parts are sampled.

Chapter 7: Confidence Intervals

281. (B) $P = 0.55$, $z^* = 1.96$ (for $C = 0.95$) $M = 0.05 = z^* \sqrt{\dfrac{p(1-p)}{n}}$

282. (D) Since σ is assumed known, we use the interval $x \pm z^* \dfrac{\sigma}{\sqrt{n}}$ where $\sigma = 81$, $n = 30$ and where z is chosen to ensure that $P(|Z| \le z) = 0.99$. From the normal tables, $P(|Z| \le 2.575) = 0.99$ (because $P(Z < 2.575) = 0.995$) and so we use $z^* = 2.575$.

283. (C) Increasing the sample size by a multiple of d divides the interval estimate by $\sqrt{d}$.

284. (C) Increasing sample size makes the hypothesis test more sensitive—more likely to reject the null hypothesis when it is, in fact, false. Increasing the significance level reduces the region of acceptance, which makes the hypothesis test more likely to reject the null hypothesis, thus increasing the power of the test. Since, by definition, power is equal to one minus beta, the power of a test will get smaller as beta gets bigger.

285. (A) Increasing sample size makes the hypothesis test more sensitive—more likely to reject the null hypothesis when it is, in fact, false. Thus it increases the power of the test. The effect size is not affected by sample size. And the probability of making a Type II error gets smaller, not bigger, as sample size increases.

286. (C)

287. (C)

288. (D)

$Z(\alpha = .05/2) = 1.96$

$2 \times 1.96 = 3.92$

289. (E) There is no guarantee that 17.8 is anywhere close to the interval.

290. (B) The margin of error $= z^* \dfrac{\sigma}{\sqrt{n}} = 0.019$. Hence the solution is 4.963 ± 0.019

291. (B) The margin of error varies directly with the critical z-value and directly with the standard deviation of the sample, but inversely with the square root of the sample size.

292. (B) Using t-scores (df $= 48$) $2.4 \pm 1.677 \left(\dfrac{0.35}{\sqrt{49}} \right) = 2.4 \pm 0.08$

293. (D) $\sigma = \sqrt{\dfrac{0.17 \times 0.83}{1700}} = 0.0091$

$z(0.0091) = 0.2$

$z = 2.20$

$0.9861 - 0.0139 = 97.2\%$

294. (C) $1.645 \times \sqrt{\dfrac{0.5^2}{n}} \le 0.04$

$\sqrt{n} \ge 20.563$

$n \ge 422.8$

Hence $n = 423$.

295. (A) Professor Miller can increase the sample size to accomplish his aims.

296. (B)

297. (D) Because there is no indication what to expect for the population proportion, we will use $p = 0.5$.

$n = \left(\dfrac{1.96}{0.05} \right)^2 = 384.16$

You would need a minimum of 385 for your sample.

298. (B) From the t-table critical value $t^* = 2.756$ with 29 df.

299. (D)

300. (B) When we estimate a standard deviation from data, we call the estimator the standard error.

301. (B)

302. (C) For a 90% confidence interval there are only two that have 1.645 as critical values. The sample size is 100. For those two combinations, the only solution is (C).

303. (A) The null hypothesis is the statement of no difference, which means H_0: $p = 0.79$. The opponent seeks evidence of a lower percent of votes, so the H_a: $p < 0.79$.

304. (B) The test statistic is a t-statistic; if we look at the t-distribution critical values for $df = 30$ and $t = 2.147$, the tail probability shows up as 0.02. Hence the p-value is 0.02.

305. (A) The critical z-values range from ± 1.645 to ± 2.326. This will result in an increase of the interval size $\dfrac{2.326}{1.645}$, which is an increase of 41%.

306. (B) The margin of error has to do with measuring variation that occurs due to chance but has nothing to do with defective survey design.

307. (A) Using the t-scores: $28 \pm 2.045 \left(\dfrac{1.2}{\sqrt{30}} \right)$.

308. (D) To divide the estimated interval by d without the confidence interval, we multiply the sample size by a multiple of d^2. In this case, $4(50) = 200$.

309. (C) Using the normal distribution tables, if $c = 0.99$, 0.99 lies between z^* and $-z^*$, 0.005 lies above z^*, which is the same as saying 0.995 is to the left of z^*. The nearest table entry shows 0.9905 as between 0.9904 and 0.9906, which corresponds to 2.345.

310. (C) $df = 4$

$$3.2 \pm 2.132 \left(\frac{0.274}{\sqrt{5}} \right) = 3.2 \pm 0.261$$

311. (B) $\sigma_d \leq \dfrac{0.5 \sqrt{2}}{\sqrt{n}}$ and $1.96 \, \sigma_d \leq 0.02$

Thus, $\dfrac{1.96(0.5)(\sqrt{2})}{\sqrt{n}} \leq 0.02$

$\sqrt{n} \geq \dfrac{1.96(0.5)(\sqrt{2})}{0.02} = 69.3$

$n \geq 69.3^2 \ 4802.5$ and hence the poll would use the sample size of 4803.

312. (D) $\sigma_p = \sqrt{\dfrac{(0.7)(0.3)}{1000}} = 0.0145$

$0.70 \pm 1.96(0.0145) = 0.7 \pm 0.0284$

$(0.672, 0.728)$

313. (B) For our problem we will use $p^* = 0.5$:

$$n \geq \left(\frac{1.96}{2(0.05)} \right)^2 = 1537$$

314. (A) $0.3 \pm 1.96\sqrt{(0.3)(0.7)/100}$

0.3 ± 0.898

$(0.210, 0.399)$

315. (A) A null hypothesis is a statement about a population parameter. It usually predicts no difference or no relationship in the population.

316. (B) We assume that a coin is fair in games of chance. When testing the statistical significance of a correlation coefficient using hypothesis testing, we assume that the correlation in the population is zero.

317. (A) When you reduce the confidence interval, the interval gets narrower.

318. (D) The claim is that the percentage is more than 12%, hence we write $p > 0.12$.

319. (E) The confidence level is not affected by the margin of error. Confidence interval is a type of interval estimate, not a type of point estimate. A population is not an example of point estimate; a sample, on the other hand, is an example of a point estimate.

320. (C) The effect size is not affected by the sample size. The probability of making a Type II error gets smaller not bigger as the sample size increases. Hence increasing the sample size makes the hypothesis test more sensitive—more likely to reject the null hypothesis. Thus it increases the power of the test.

321. (A) Any confidence interval is found by adding and subtracting the margin of error from the point estimate. Therefore any confidence interval is 2E across where E is the margin of error. The margin of error can be found by subtracting the endpoints and dividing by 2.

322. (C) Before the confidence interval can be calculated, the point estimate must be calculated. In this case it is $0.255 = 37/145$. Once this is found, the confidence interval can be calculated using the formula $0.255 \pm 1.96\sqrt{\dfrac{0.255(1-0.255)}{145}}$.

323. (D) The sample mean will be halfway between the two endpoints of the confidence interval. Therefore we can find the sample mean by averaging the lower and upper bounds of the confidence interval: $(12.9 + 18.6)/2 = 15.75$.

324. (A) Mathematically, increasing the sample size will decrease the value of the margin of error since the sample size is in the denominator of the formula.

325. She is estimating a proportion. Therefore the minimum sample size formula for estimating proportions should be used. Since we do not have a preliminary estimate, we will use 0.5 as an estimate.

$$0.5(1-0.5)\left(\frac{1.96}{0.02}\right)^2 = 2401$$

326. **(a)** $0.75 \pm 2.576\sqrt{\dfrac{(0.75)(0.25)}{1000}} = 0.7147$ to 0.7853

 (b) $n = (0.75)(0.25)(2.576/0.02)^2 = 3110.52$, rounded up to 3111.

327. **(a)** Let $p_1 =$ the proportion of the home games won and $p_2 =$ the proportion of the away games won.

 (b) The standard error for a confidence interval estimate of $p_1 - p_2$ is

$$\sqrt{\dfrac{(0.593)(0.407)}{81} + \dfrac{(0.543)(0.457)}{81}} = 0.0777.$$

The count of successes and failures is at least five, so the z-procedures should be fairly accurate. Our 90% confidence interval is $(p_1 - p_2) \pm z^* SE = (0.593 - 0.543) \pm 1.645 (0.0777) = (-0.078, 0.178)$. This confidence interval means that there is a 90% likelihood that the true difference between the proportion of games won at home compared to games won away is between -0.078 and 0.178. Since zero lies in this confidence interval, we are less than 90% sure that the true proportion of games won at home is greater than the true proportion of games won away.

328. **(a)** With a df $= 9$,

$$\bar{x} \pm t\dfrac{s}{\sqrt{n}} = 800{,}000 \pm 2.262\,\dfrac{315{,}000}{\sqrt{10}} = 800{,}000 \pm 225{,}000.$$

 (b) We assume the sample is an SRS and that the baseball salaries are normally distributed.

329. **(a)** If a Type I error was committed, we would conclude more than 1% were defective when they really were not. The shipment would be returned when it should not have been returned.

 If a Type II error was committed, we would conclude that there were no more than 1% of the panels defective when they really were. The faulty panels would be accepted from the supplier.

 (b) The supplier would think that Type I error is more serious because they would receive the panels back even when they are not defective.

 (c) The TV manufacturer would think Type II error is more serious because the accepted panels will be of poor quality.

330. Since the confidence interval is 0.95, the z-value is 1.96.

 (a) $95 - 90.5 \pm (1.96)\sqrt{\dfrac{(4.5)^2}{300} + \dfrac{(4.1)^2}{250}}$

 4.5 ± 0.719

 (b) $4.5 \pm 0.719 -\!/(3.78, 5.22)$

 (c) There is a 95% likelihood that the true difference in average test score between U.S. and Canadian ninth-graders is between 3.78 and 5.22. Since zero does not lie in this confidence interval we are more than 95% sure that the U.S. test score average is higher than the Canadian test score average.

Chapter 8: Inference for Means and Proportions

331. (C) There are two independent random samples taken and the hypothesis being tested involves the mean.

332. (D) There are two independent samples taken and the hypothesis being tested involves the population proportion.

333. (D) There is one random sample and the hypothesis being tested involves the mean.

334. (B) The 45% is stated as a population proportion to which we are comparing a single sample.

335. (D) The average 2 years ago is stated as a population mean, which we are comparing a single sample to.

336. (C) We are comparing the means of two independent random samples.

337. (C) Since the null hypothesis was not rejected, the p-value is larger than 0.05 and may or may not be less than 0.10. Further, the sample mean may or may not be 12.5.

338. (A) If Statement I is true, then the sampling distribution of the sample proportion is approximately normal and z-procedures can be used.

339. (B) A two-sided test is of the form H_0: $\mu = c$ versus H_a: $\mu \neq c$ for a constant c. If the null hypothesis is rejected, the 90% confidence interval will not contain c.

340. (B) If the null hypothesis is not rejected, the confidence interval must contain 7. In this case, only the confidence interval in (B) contains that value.

341. (A) Since there are two independent samples being compared, the test statistic should be $t = \dfrac{\bar{x}_1 - \bar{x}_2}{\sqrt{\dfrac{s_1^2}{n_1} + \dfrac{s_2^2}{n_2}}}$.

342. (E) It is important to understand that you can't base a decision about the population on the sample statistic alone. Here we would need the standard deviation to be able to be able to perform the test using either a test statistic or a p-value.

343. (E) Without the sample size, the test statistic cannot be calculated and therefore the test cannot be performed.

344. (A) The given hypothesis is tested with a left-tailed t-test and the p-value can be found using the t-test on the TI calculator or by finding the probability that the sample mean would be less than 43, given the population mean is 45 and using the standard deviation for the sampling distribution.

345. (C) Using the t-table, the area in the tail of the t-distribution associated with this test statistic and 19 degrees of freedom is between 0.05 and 0.10. Since this is a two-sided test, the p-value would be between 0.10 and 0.20.

346. (D) Since 156 is contained in this confidence interval, it cannot be excluded as a possible value of the true mean.

347. (B) If we performed a two-sided hypothesis test, the null hypothesis would be rejected since 0.3 is not contained in the interval.

348. (C) More than half is equivalent to the statement that the population proportion is more than 0.5.

349. (E) Since this is a test of the population mean, choice (D) can be eliminated. Further, the statement "the mean is larger than 35,000" is equivalent to the statement $\mu > 35,000$.

350. (A) A paired samples t-test is used to test the difference of the means of two dependent samples. In this type of test, the t-distribution is used to find the p-value using the mean difference as a point estimate.

351. (C) The formula for a one-sample t-test of the mean is $t = \dfrac{\overline{x} - \mu}{s/\sqrt{n}} = \dfrac{\overline{x} - 0}{2.16/\sqrt{20}}$.

Solving for the sample mean algebraically gives 0.6033.

352. (A) Since we are testing the proportion of a single population, the test statistic is

$$z = \frac{\hat{p} - p}{\sqrt{\dfrac{p(1-p)}{n}}} = \frac{0.6 - 0.65}{\sqrt{\dfrac{0.65(0.35)}{390}}} = -2.070.$$

353. (B) While the sample proportion can be used to find the test statistic and then the p-value, by itself it cannot be used to make a determination about the population proportion. Since the null hypothesis was rejected at the 10% level, it must be that the p-value is less than 0.10. Further, it was not rejected at the 5% level, so the p-value must be larger than 0.05.

354. (D) When testing for a difference of means with two dependent samples, the individual samples aren't used but instead the set of differences is used.

355. (A) A pooled estimate is used in two-sample t-tests when there is evidence that suggests the two population variances are equal. It should be noted that this rarely occurs with actual data.

356. (C) If the p-value for the one-sided test is 0.04711, then the p-value for the two-sided test will be double this value.

357. (D) Since the p-value for the left-tailed test is 0.0625, the p-value for the right-tailed test will be $1 - 0.0625$.

358. (B) We are testing two proportions from two independent samples, therefore the test statistic is

$$z = \frac{\hat{p}_1 - \hat{p}_2}{\sqrt{\left(\frac{X_1+X_2}{n_1+n_2}\right)\left(1-\frac{X_1+X_2}{n_1+n_2}\right)\left(\frac{1}{n_1}+\frac{1}{n_2}\right)}} = \frac{\frac{312}{500} - \frac{400}{525}}{\sqrt{\left(\frac{312+400}{500+525}\right)\left(1-\frac{312+400}{500+525}\right)\left(\frac{1}{500}+\frac{1}{525}\right)}}.$$

This can also be found by performing a "2-prop-z-test" on the TI84 or 83 calculator as it gives both the test statistic and the p-value.

359. (C) Performing the test with the alternative $p < 0.3$ yields a p-value that is less than 0.10. Therefore the null hypothesis will be rejected, suggesting that the proportion is in fact less than 30%.

360. (A) If all of the sample statistics such as the mean and standard deviation are the same then an increase in the sample size will decrease the denominator of the test statistic, making the overall test statistic larger. This larger test statistic will leave less area in the tail for the p-value.

361. (A) Since the p-value is less than 0.05, the null hypothesis will be rejected and the statement of the alternative will be supported.

362. (B) With a sample of 20, there are 19 degrees of freedom associated with this test, giving a critical value of -2.539. Our test statistic is larger, so the null hypothesis is not rejected and the statement of the alternative is not supported.

363. (B) Since the p-value is larger than 0.05, the null hypothesis is not rejected and the statement of the alternative cannot be supported.

364. (A) The critical value for this test is 1.645. Since $2.325 > 1.645$, the null hypothesis is rejected and the statement of the alternative is supported.

365. (E) In order to perform the test, we need to calculate the critical value. In order to find this value we must determine the degrees of freedom which depend on the sample sizes.

366. (C) Since the p-value is less than 0.10, the null hypothesis is rejected and the statement of the alternative is supported.

367. (C) The described samples are dependent since they are measurements on the same group of people, therefore a paired samples t-test should be used. Further, since the p-value is larger than 0.05, there is no evidence the reaction time has increased.

368. (C) Since the samples are measurements on the same group, the paired samples t-test should be used. If the speeds are increased, the runners' times will decrease and by letting the times before represent Sample 1, we will have the alternative $\mu_d > 0$. With 15 degrees of freedom, the critical value is 2.602, and since our test statistic is smaller we do not reject the null hypothesis and cannot support the statement of the alternative.

369. (B) This is a two-sample proportion test and the p-value can be calculated using the TI83 or 84 as 0.06345 (the test statistic is -1.526). Since this is larger than 0.05, the null hypothesis is not rejected.

370. (A) Since the test statistic comes from the standard normal distribution, this is the same distribution used to find any p-value associated with this type of test. Samples are assumed to be independent but the sample sizes do not need to be the same.

371. (B) Proportion is not relevant to this problem, so (C) is incorrect. Choice (B) is the only appropriate answer because "not equal to 290,000" contradicts the realtor's statement.

372. (C) I and III are correct from their definitions. The p-value does not give the probability that the null hypothesis is true. The p-value of a sample statistic is the probability of obtaining a result as extreme as the one obtained.

373. (B) I is correct and II is incorrect from the definitions of null and alterative hypotheses. The p-value of a sample statistic is the probability of obtaining a result as extreme as the one obtained: the higher it is, the more likely is the null hypothesis to be true.

374. (E) The rule of thumb is 30 not 50.

375. (D) I is correct. When doing inference for a population mean we usually use t-procedures rather than z-procedures because the true population standard deviation is unknown. (II and III) z-procedures require a sample drawn from a normally distributed sample, not a binomially distributed sample.

376. (D) Definition of one-sided alternative hypothesis.

377. (C) We usually use t-procedures with inference for a population mean because z-procedures assume that we know the population's standard deviation, which we seldom do. z-procedures are in fact often appropriate for proportions with large samples. The definition of robust makes III correct.

378. (C) I and III are correct by the definitions of null hypothesis and Type I error. II is incorrect because you can argue your conclusion based on the p-value alone: if it is small you have evidence against H_0.

379. (C) 0.081. $H_0 : \mu = 64 \ H_A : \mu > 64$. The standard deviation of sample means is

$$\sigma_{\bar{x}} = \frac{\sigma}{\sqrt{n}} = \frac{.5}{\sqrt{49}} = .0714. \text{ The } z\text{-score for 64.1 is } \frac{64.1 - 64}{.0714} = 1.4. \text{ Using Table A, } P(z < 1.4) =$$

.9192. Therefore if the 64-ounce claim is correct, the probability of obtaining a sample mean greater than 64.1 and mistakenly rejecting the claim is $1 - .9192 = .081$.

380. (C) The t-score of 131 is $\dfrac{131-137}{20/\sqrt{32}} = -1.697$. $p = .0499$.

On the TI-84, STAT → TESTS → T-Test, then Inpt:Stats, μ_0: 137, $\bar{x}$: 131, Sx: 20, n: 32, Calculate yields $t = -1.697056$ and $P = .04985$.

381. (E) A paired samples test is performed when the samples taken are dependent or matched in some way, so a two-sample t-test would not be appropriate. Only I and III represent statements which must be true.

382. (A) The p-value of any test represents the probability of obtaining a sample as extreme or more extreme than the sample taken if the equality statement of the null hypothesis is true.

383. (B) If only n is increased, the sample proportion will decrease, leading to a less probable sample under the null hypothesis (although the null may or may not be rejected). This could also be thought of in terms of the standardized test statistic. If the sample proportion is decreased, the standardized test statistic will also decrease.

384. (A) When the null hypothesis is not rejected, we cannot say whether the statement is true or not, only that there is not enough evidence to convince us of the alternative statement. Further, the level of significance is actually the probability of rejection of the null hypothesis if the null hypothesis is true, a Type I error.

385. (A) Since this is a right-tailed test, the null will be rejected if the standardized test statistic is larger than the critical value of 1.645.

386. (a) For Neighborhood 1, the point estimate is $258/392 = 0.6582$; for Neighborhood 2, the point estimate is $481/517 = 0.9304$.
   (b) To determine if there is a statistically significant difference, a test of H_0: $p_1 = p_2$ versus H_a: $p_1 \neq p_2$ should be performed. This two-proportion z-test yields a p-value of approximately zero, which is strong evidence to reject the null hypothesis at any significance level. Therefore there is strong evidence of a difference in the proportions.

387. (a) Since the organization is interested in claiming that the mean is larger than 200, the test performed should be H_0: $\mu \leq 200$ versus H_a: $\mu > 200$.
   (b) The p-value of any test represents the probability of obtaining a sample as extreme or more extreme than the sample taken if the null hypothesis is true. In this case, there is a 1.65% chance of obtaining a sample with a mean of 215.67 or more if the true mean is in fact 200.

388. (a) The interval can be calculated using the formula for a one-proportion confidence interval or by using 1-prop Z interval on the TI83/84. The resulting interval is (0.562, 0.656).
   (b) This null hypothesis would not be rejected since 0.61 is in the 95% confidence interval.

389. **(a)** There are many possible factors other than the supplement that could affect this study. One example is the fact that the subjects took the test twice and may perform better the second time just from practice even though the words are different. Also, some individuals may naturally have better memory than others, although this should be taken care of by the pairing of the samples and the randomization.

    (b) To answer this question a paired samples t-test must be performed with the hypotheses $H_0: \mu_d \geq 0$ versus $H_a: \mu_d < 0$. The resulting p-value is 0.1352, which is not significant at the 5% or 10% levels. Therefore there is not convincing evidence that the number of words recalled was increased.

390. **(a)** Since the p-value for the two-sided test is 0.558, the p-value for the one-sided test will be 0.279, which is not less than 0.05. Therefore there is not convincing evidence that the mean is larger than 4.5.

    (b) The given p-value is based on assuming that the equality statement of the null hypothesis is true, that is that the true mean is 4.5. Therefore this p-value will not give us any information for a hypothesis test using a hypothesized value of the true mean of 4.7.

Chapter 9: Inference for Regression

391. **(A)** With such a small p-value, the null hypothesis will be rejected at the 5% level of significance. The evidence suggests there is a nonzero slope and thus a linear relationship likely exists between the variables.

392. **(A)** Based on the computer output, the p-value is zero, which would reject the null hypothesis for all of the usual significance levels. This is strong evidence of a linear relationship between the variables.

393. **(E)** Testing whether the slope is nonzero is equivalent to testing whether the population correlation coefficient is nonzero.

394. **(B)** The value of the y-intercept is not determined by the strength, direction, or even the presence of a linear relationship between two variables.

395. **(B)** The test statistic for any test is always of the form $\dfrac{\text{estimate-parameter}}{\text{standard error of the estimate}}$. Since we are testing whether the slope is zero, the test statistic is $\dfrac{0.09}{0.76} = 0.12$.

396. **(C)** Since the 95% confidence interval contains zero, it is possible at the 5% level of significance that the slope is zero.

397. **(A)** A positive linear relationship would yield a positive slope.

398. **(A)** A negative linear relationship would yield a negative slope.

399. (E) To find the test statistic, we would need the estimate of the slope, 1.04, and the standard error of the estimate. Because the *t*-critical value depends on the number of (x, y) data pairs, it isn't possible to find s_b algebraically using the margin of error without knowing the sample size.

400. (A) The *t*-test statistic is $\dfrac{1.51}{s_b}$. Using algebra, $s_b = 0.21$.

401. (D) Since the regression equation is listed as "Height = 5.31 + 3.34 Percentage," it is clear that percentage is the independent variable. Therefore on the computer output, the line "Percentage" lists information for the slope. Using this line shows that both I and II are true. The last statement is not true, as the standard error of the slope is 1.316.

402. (E) Although it is clear that the slope of the regression line is positive, we cannot conclude that this is true for the population without more information.

403. (A) Since "square feet" is used as the independent variable, that line provides information about the slope. Thus, the *t*-test statistic is 1.53.

404. (C) With the given information the *t*-test statistic can be calculated as 1.293/2.967 = 0.4358. The critical values for a two-sided test with a significance level of 0.05 are −2.179 and 2.179 (with 12 degrees of freedom), leading us not to reject the null hypothesis.

405. (A) When the null hypothesis is rejected, the evidence suggests that the true regression line has a nonzero slope, which is equivalent to there being a linear relationship between the variables.

406. (E) In this test, when the null hypothesis is not rejected, there is simply not enough evidence to suggest the alternative may be true.

407. (A) By definition, a confidence interval gives the possible bounds on the population parameter, which in this case is the true slope of the regression line. Although the statement of (B) seems like it should be true, it actually isn't since the true slope is a fixed value and it does not make sense to talk about a probability associated with it.

408. (A) The slope of any regression line can be interpreted as the average change in y for a 1-unit increase in x. Since this is a 90% confidence interval, we get a range of possible values for this rate of change.

409. (D) Letters A–C are all possible since the interval contains positive values, negative values, and zero. Similarly, the interval contains 2.11, making (E) possible as well. It is because zero is present in the interval that the null hypothesis stated in choice (D) will not be rejected.

410. (B) Every confidence interval has the form (point estimate) ± (critical value)(standard error). Therefore $(t)(s_b) = 0.981$. The critical value can be found using the *t*-table with 23 degrees of freedom and is 2.069.

411. (C) Every confidence interval has the form (point estimate) $\pm$ (critical value)(standard error). The t critical value with 19 degrees of freedom is 1.729, giving an interval of $-1.215 \pm (0.8496)(1.729)$.

412. (B) If the null hypothesis is not rejected, it is possible that the true slope is zero. A confidence interval calculated from the same sample data would therefore need to contain zero.

413. (A) If the null hypothesis is rejected in this test, it suggests the true slope is positive. A confidence interval calculated using the same sample data should therefore contain only positive values.

414. (C) If the null hypothesis is rejected in this test, it suggests the true slope is negative. A confidence interval calculated using the same sample data should therefore contain only negative values.

415. (B) The confidence interval does not contain zero, so a hypothesis test using the same data would reject the null hypothesis. In this type of test, if the null hypothesis is rejected, it suggests a probable linear relationship in the data.

416. (C) The confidence interval contains all positive values, therefore a test of this type using the same dataset would lead to not rejecting the null hypothesis.

417. (B) The confidence interval contains all positive values, therefore a test of this type using the same dataset would lead to rejecting the null hypothesis, concluding that there is evidence suggesting a positive linear relationship between the two variables.

418. (D) This question is essentially asking us to find the point estimate of the slope, which is at the center of the given confidence interval. The given confidence interval is for the slope since the interpretation is one used for slope. To find the center of the confidence interval add the endpoints and divide by 2 to get 3.221.

419. (C) A 99% interval is equivalent to using a significance level of 1% on this two-sided test. Since $0.025 > 0.01$, we would not reject the null hypothesis and conclude that it is possible the slope is zero. Therefore the confidence interval would contain zero.

420. (E) To find the critical value to compare with the t-test statistic, we would need the sample size, which is not provided in the question.

421. (E) The fact that the regression is significant only tells us that the slope is likely to be nonzero.

422. (A) The statement "test for significance of regression" refers to determining whether the slope of the regression line is nonzero (meaning the correlation will also be nonzero).

423. (A) Since the p-value is less than 5%, the null hypothesis in a test of H_0: $\beta = 0$ versus H_a: $\beta \neq 0$ will be rejected, suggesting that there is a nonzero slope in the population regression line. These data cannot be used to make a determination about the y-intercept or the quality of any predictions made from the line.

424. (B) To determine whether the slope is positive, the test used would be H_0: $\beta \leq 0$ versus H_a: $\beta > 0$.

425. (E) The given slope of 3.197 is a point estimate. Without more information such as the standard error of that estimate, the hypothesis test cannot be conducted.

426. (D) Since the point estimate of the slope is negative and any confidence interval will be centered around it, the confidence interval will contain some or all negative values.

427. (C) If the confidence interval contained zero, the null hypothesis would not have been rejected since that would imply zero is a possible slope at the 5% level.

428. (D) The t-test statistic to perform this test is $t = -2.592/1.91 = -1.357$, and the critical value is 1.771. Since the absolute value of the test statistic is less than 1.771, the two-tailed 10% t-value for df $= 13$ and the null hypothesis is not rejected.

429. (D) The t-test statistic to perform this test is $0.0328/0.072 = 0.4556$ and the critical value is 2.101. Since $t < 1.735$, the one-tailed 5% t-value for df $= 18$ and the null hypothesis is not rejected.

430. (D) The t-test statistic to perform this test is $-3.91/0.84 = -4.655$, and the critical one-tailed t-value for df $= 16$ is -1.746. Since $t < -1.746$, the null hypothesis is rejected.

431. (A) The 95% confidence interval will be of the form (point estimate) $\pm$ (critical value) $\times$ (s_b). Using the formula for the test statistic, $s_b = 2.618$. With a sample of 20, the critical value is 2.093, giving a confidence interval of $-1.475 \pm (0.831)(2.093)$.

432. (C) The confidence interval estimate of the true slope of the regression line has no relationship to the y-intercept of the regression line.

433. (A) Before answering this question, note that choice (D) can be eliminated since a p-value cannot be larger than 1. If the test determined there was a linear relationship, the null hypothesis of the test must have been rejected. Therefore the p-value must be less than 0.10.

434. (D) If the test determined that there is evidence of a linear relationship, then the null hypothesis was rejected. In other words, the t-test statistic is greater than the critical value of 1.725.

435. (B) In general, the formula for the t-test statistic of this type of test is $\dfrac{\text{estimate of the slope}}{s_b}$. Solving algebraically for the estimate of the slope shows the regression line must have a slope of 3.024.

436. (B) Any confidence interval has width 2E where E is the margin of error and has the point estimate at its center. Therefore the point estimate of the slope is $\dfrac{-3.378 + (-0.936)}{2}$.

437. (B) The predictor, or independent variable, in this situation is "Hours." On the computer output, the *p*-value for this test can be found on the line for "Hours."

438. (B) The predictor, or independent variable, in this situation is "Hours." On the computer output, the *t*-test statistic for this test can be found on the line for "Hours."

439. (D) The critical value to find this interval is 2.306. From the output, the standard error of the estimate is 0.2605. Thus, the margin of error is (2.306)(0.2605).

440. (a) Reading the computer output gives $Y = 59.603 - 0.23335X$ where Y is the time spent on the assessment and X is the number of practice problems completed in the week before the assessment.
   (b) To test whether there is a linear relationship between the variables, the null and alternative H_0: $\beta = 0$ versus H_a: $\beta \neq 0$ should be used. Since this test involves the slope, the *p*-value 0.020 should be used. In this case, at the 5% level we would reject the null hypothesis and conclude that there is a linear relationship between the number of practice problems completed and the time spent on the assessment.

441. (a) Since the slope is positive, there would be a positive/direct relationship between the two variables in the sample.
   (b) There is not enough information to calculate the interval. At a minimum, we would need either the sample size or degrees of freedom associated with the estimate and the standard error of the estimate.

442. (a) Since the independent variable is sugar, we use the information on that line of the computer output. A 95% confidence interval would therefore be 0.14011 ± (0.07797)(2.080) where 2.080 is the critical value associated with this level of confidence and 21 degrees of freedom.
       The interpretation provides a description of the slope in terms of the data being modeled: we are 95% confident that every one-gram increase in sugar will result in an average change in price of between −$0.02 and $0.30.
   (b) The regression would be found not to be significant since the confidence interval contains zero.

443. (a) The linear relationship in the population could be positive, negative, or nonexistent since the confidence interval contains zero, positive, and negative values.
   (b) A 99% confidence interval will always be wider than a 95% confidence interval, so no, it would not change the description in (a).

444. (a) Rejecting the null hypothesis in this problem implies that there is a linear relationship between the time after the drug is administered and the time until the virus is no longer present.
   (b) Since the null hypothesis was rejected at the 10% level, a 90% confidence interval calculated from this dataset would not contain zero.

445. (a) Since the slope of the regression line is negative, there will be a general negative correlation between the variables seen in the scatter plot.

(b) The only statement that can be made about the 95% confidence interval is that the sample slope -1.253 will be at the center. The width of the confidence interval cannot be determined from the information provided.

Chapter 10: Inference for Categorical Data

446. (A) This is a property of the chi-squared distribution. All values covered by this distribution are positive.

447. (C) The degrees of freedom for a chi-squared goodness of fit test are the number of categories $n - 1$.

448. (C) A contingency table organizes the observed frequencies such that each cell represents the frequency for the joint value of the row and column variables. This information is then compared to the expected frequency using a chi-squared test to determine whether the variables are related.

449. (A) For a chi-squared test of independence, the null hypothesis is that the variables are independent and the expected frequencies are calculated under this assumption. Therefore if the observed and expected frequencies differ significantly, there is less evidence that the null hypothesis holds.

450. (D) (Look up in the table for chi-squared distribution, df $= 4 - 1, 0.05 = 7.81$.)

451. (D) The degrees of freedom of the chi-squared distribution is equal to the number of standard normal deviates being summed, which is 14 in this case.

452. (A) The mean of chi-squared distribution is its degrees of freedom.

453. (D) As the degrees of freedom of a chi-squared distribution increase, the chi-squared distribution begins to look more like a normal distribution. In these sets of choices the 10 df would look the most similar to the normal distribution.

454. (B) The categorical variable in this problem is the grade distribution. The different population is the four teachers. The test is to determine whether their grades are homogeneous.

455. (B) If $n = 14$, the df $= 14 - 1 = 13$. In the chi-squared table with the probability (right-tail area) of 0.05 and the row with 13 df, we find the value to be 22.36.

456. (A) The chi-squared goodness of fit test is used to find out whether an observed pattern of categorical data is consistent with a specified distribution. In this problem, we are dealing with a categorical variable—the type of baseball card. And we want to determine whether the distribution in our package is consistent with the production distribution claimed by the local store. So the chi-square goodness of fit test is appropriate.

457. **(A)** Using the formula for the expected frequency of this cell,

$$\frac{(\text{total male})(\text{total independent})}{\text{overall total}} = \frac{(400)(100)}{1,000} = 40.$$

458. **(D)** In a chi-squared test of independence, we compare the observed and expected frequencies to determine whether two variables are related.

459. **(A)** For a two-tailed test of this type, it will be possible to write the alternative in the form of parameter ≠0.

460. **(D)** With a 2×2 design, the chi-squared test is similar to a z-test for proportions and directional hypotheses make sense. With more categories, there are more proportions to compare and this no longer works.

461. **(E)** Chi-squared distributions cannot have negative values.

462. **(C)** A contingency table is sometimes referred to as a cross-classification table or in some cases a two-way table.

463. **(B)** The expected counts represent the frequency we would expect to see if the null hypothesis were true.

464. **(C)** While possible, combining all of the levels of one category would possibly affect our analysis. For instance if we were interested in gender differences, combining male and female would no longer allow us to do this. We could, however, combine the team and contact sport categories since these are the most related and still perform a similar analysis.

465. **(E)** Although it may indirectly involve one or more of these values, chi-squared tests are used to test for independence between two variables and whether a particular distribution "fits" the observed data.

466. **(C)** The standard deviation of the population is 4 minutes. The standard deviation of the sample is 12 minutes. The sample is 200. $\chi^2 = \frac{[(n-1)^2 \times s^2]}{\sigma^2}$ $\chi^2 = \frac{[(7-1)^2 \times 6^2]}{4^2} = 81.$

467. **(B)** The expected two-way table is given by the matrix $\begin{bmatrix} 7.5 & 20 \\ 7.5 & 20 \end{bmatrix}$

$$\chi^2 = \frac{(5-7.5)^2}{7.5} + \frac{(25-20)^2}{20} + \frac{(10-7.5)^2}{7.5} + \frac{(15-20)^2}{20} = 4.167$$

468. **(A)** Degrees of freedom = (rows − 1)(columns − 1) = (4 − 1)(2 − 1) = 3.

469. **(D)** The p-value for the χ^2 statistic is 0.1865, which is more than any reasonable alpha value. We would fail to reject the null hypothesis. There is insufficient evidence to conclude that an association or relationship exists between the weight of the individual and the weight he can lift.

470. (B) The correct hypothesis of the slope of a regression line is the one-sample t-test. Since you are dealing with sample data, you do not have the population values needed to use a z-test. Although you have two variables—one dependent and one independent—you only have one sample.

471. (A) The t-test for the slope of regression line is testing whether there is a relationship between x and y. If a relationship exists, the slope will not be 0. Therefore the null hypothesis is that the slope is zero; the alternative hypothesis is that the slope is not zero.

472. (C) We have $2 + 3 + 3 + 4 + 6 = 18$, and thus the expected numbers according to the superintendent's claim are

$$\frac{2}{18}(6000) = 667, \frac{3}{18}(6000) = 1000$$

$$\frac{4}{18}(6000) = 1333, \frac{6}{18}(6000) = 2000.$$

Using chi-square to test for goodness of fit, we calculate

$$\chi^2 = \frac{(720-667)^2}{667} + \frac{(970-1000)^2}{1000} + \frac{(1013-1000)^2}{1000} + \frac{(1380-1333)^2}{1333}$$
$$+ \frac{(1917-2000)^2}{2000} = 10.38.$$

With df $= 5 - 1 = 4$, the p-value is $p = (\chi^2 > 10.38) = 0.0345$. Note that $0.0345 < 0.05$, but $0.0345 > 0.025$. Thus, there is sufficient evidence to reject the superintendent's claim at the 5% significance level but not at the 2.5% level.

473. (C) Here we have four different populations: residents of New Jersey, Washington, Texas, and Georgia. The economist then compares the distribution of income for these four populations. Thus, we have one categorical variable—distribution of income—and four populations.

474. (C) Degrees of freedom $=$ (rows $- 1$)(columns $- 1$) $= (3 - 1)(4 - 1) = 6$.

475. (A)

$$\frac{897 + 905 + 967}{3} = 923$$

$$\chi^2 = \frac{(897-923)^2}{923} + \frac{(905-923)^2}{923} + \frac{(967-923)^2}{923} = 3.1809$$

With df $= 3$, the critical χ^2-values for 0.05, 0.01, and 0.001 are 5.99, 9.21, and 13.82, respectively. Since 3.18 is less than each of those, there is sufficient evidence at all of these levels to say that the number of accidents on each shift is not the same.

476. (B) $p(novice) = \dfrac{120}{200}$

477. (A) $p(uphill) = \dfrac{30}{200}$

478. (D) $p(novice\ and\ uphill) = p(N) \times p(U) = \left(\dfrac{120}{200}\right)\left(\dfrac{30}{200}\right) = \dfrac{9}{100}$

479. (D) We have H_0: good fit with a 1:1:2:2 ratio. Noting that $45 + 33 + 74 + 76 = 228$, $\left(\dfrac{1}{6}\right)228 = 38$, and $\left(\dfrac{2}{6}\right)228 = 76$. The χ^2 value for these data (degrees of freedom $= 3$) is about 3.66.

480. (C) χ^2 test for homogeneity because the categorical variable is the grade distribution. The three professors are three different populations. The test is to determine whether their grade distributions are homogeneous.

481. (D) The entire curve of the χ^2 distribution lies to the right of the y-axis and thus assumes only nonnegative values.

482. (B) The correct test for the hypothesis of the slope of a regression line is the one-sample t-test. Since you are dealing with sample data, you do not have the population values needed to use a z-test. Although you have two variables—one dependent and one independent—you only have one sample.

483. (C) There are three assumptions/conditions that must be checked in order to proceed with the chi-squared test for goodness of fit, and one of them is that all expected cell counts are at least 5.

484. (C) $df = (r - 1)(c - 1) = (3 - 1)(4 - 1) = 6$

485. (E) The χ^2 distribution is strongly skewed to the right for small degrees of freedom and becomes closer to symmetric (bellshaped) as the degrees of freedom get larger.

486. (B) A specific chi-squared distribution is specified by one parameter, called the *degrees of freedom*.

487. (D) Since there are 12 months in a year, $n = 12$ and $df = 12 - 1 = 11$. From the table, $0.01 < p < 0.02$.

488. (D) In this case $n = 10$, $df = 10 - 1 = 9$. The associated χ^2 is 21.67.

489. (C) There isn't enough information for us to reject the claim. The χ^2-value is about 5.16, which is not large enough to reject the null hypothesis.

490. (A) The chi-squared goodness of fit test is used to find out whether an observed pattern of categorical data is consistent with a specified distribution. In this problem, we are dealing with a categorical variable—the type of baseball card. And we want to determine whether the distribution in our package is consistent with the production distribution claimed by ABC Toy Company. So the chi-squared goodness of fit test is appropriate. The other chi-squared options (the independence test and the homogeneity test) involve comparing data from two samples. Since we only have one sample in this baseball card problem, they are not appropriate. The t-tests are not appropriate since they are used with quantitative data, and this problem involves categorical data.

491. (D) To determine whether two variables are independent, we always use a chi-squared test of independence.

492. (D) To determine whether the two variables are related, we will use a chi-squared test of independence and get a test statistic of 2.78, which leads us to fail to reject the null hypothesis that the two variables are independent.

493. (A) The requirement on the expected frequencies given by most textbooks is 5, not 30. Although the data are often presented in a table, this is not a requirement for performing the test.

494. (D) Cell 1,3 represents the value in the first row and third column. The expected frequency of this cell is $\left[\dfrac{(110+120+115)(115+101+88)}{899} \right]$.

495. (a)

	1	2	3	4	5
Full Time	27	55	45	12	30
Part Time	10	12	15	27	11

(b) To test this claim we use a chi-squared test of independence with the null hypothesis being that the two variables are independent and the alternative being that they are dependent. This results in a test statistic of 33.78, which leads us to reject the null hypothesis (the p-value is very nearly zero). Therefore there is evidence that the variables are related.

496. (a) $0.3(500) = 150$, $0.25(500)$ 125, $0.2(500) = 100$, $0.25(500) = 125$, so we have

	Channel			
	2	3	4	5
Expected Number	150	125	100	125

(b) We note that the expected values (150,125,100,125) are all > 5.
H_0: The TV audience is distributed over Channels 2, 3, 4, and 5 with percentages 30%, 25%, 20%, and 25%, respectively.

H_a: the audience distribution is not 30%, 25, 20%, and 25%, respectively
We calculate the chi-square:

$$\chi^2 = \frac{(139-150)^2}{150} + \frac{(138-125)^2}{125} + \frac{(112-100)^2}{100} + \frac{(111-125)^2}{125} = 5.167$$

To use the χ^2 at a 10% significance level with df = 3, the critical χ^2-value is 6.25. Since $5.167 < 6.25$, there is not significant evidence to reject H_0.

497. (a) Goodness of Fit Test – Equal Proportions. Since there are five groups, the probability is $1/5 = 0.2$. Since there are 500 numbers generated, expected counts are $500(0.2) = 100$. The formula for $\chi^2 = \Sigma \dfrac{(O-E)^2}{E}$. For these data, the calculation is

$$\frac{(103-100)^2}{100} + \frac{(90-100)^2}{100} + \frac{(110-100)^2}{100} + \frac{(89-100)^2}{100} + \frac{(95-100)^2}{100} = 3.55.$$

There are four degrees of freedom here, so the p-value is 0.3041.

Conclusion: Since the p-value is high ($p > .05$), we fail to reject the null hypothesis. So we can conclude that the random numbers are uniformly distributed among the five groups.

498. Contingency Table

H_0: There is no relationship between income and whether the person played the lottery.
H_a: There is a relationship between income and whether the person played the lottery.
Reject H_0 if χ^2 is greater than 5.991.

(c) $\chi^2 = \dfrac{(46-40.71)^2}{40.71} + \dfrac{(28-27.14)^2}{27.14} + \dfrac{(21-27.14)^2}{27.14} = 6.54$

Reject H_0. There is a relationship between income level and playing the lottery.

499. (a) H_0: Gender and voting preferences are independent.
H_a: Gender and voting preferences are not independent.
(b) df $= (2-1)(3-1) = 2$

$$\chi^2 = \frac{(200-180)^2}{180} + \frac{(150-180)^2}{180} + \frac{(50-40)^2}{40} + \frac{(250-270)^2}{270}$$

$$+ \frac{(300-270)^2}{270} + \frac{(50-60)^2}{60} = 16.2$$

The p-value is the probability that a chi-squared statistic having two degrees of freedom is more extreme than 16.2.

We use the Chi-Square Distribution Calculator to find $p(\chi^2 > 16.2) = 0.0003$. Since the p-value (0.0003) is less than the significance level (0.05), we cannot accept the null hypothesis. Thus, we conclude that there is a relationship between gender and voting preference.

500. (a) p (less than 250,000 newspapers sold) $= 0.01 + 0.07 + 0.14 + 0.25 = 0.47$.

 (b) The expected amount is $\$100,000 + \$10,000 \cdot (0.15 + 0.1 + 0.04) = \$100,000 + \$2,900 = \$102,900$.

 (c) $p((x > 260,000)$ and $(270,000 < x < 280,000))$
$= p(270,000 < x < 280,000)/p(x > 260,000)$
$= 0.1/(0.15 + 0.1 + 0.04) = 0.345$.